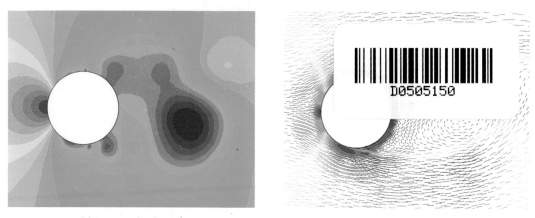

(a) Pressure distribution   (b) Velocity vectors

Plate 1 Flow around a circular cylinder (finite volume method), $Re = 100\ 000$ (page 264)

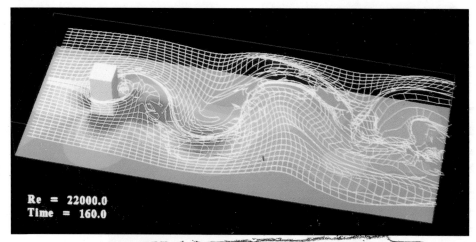

Re = 22000.0
Time = 160.0

Plate 2 Turbulent flow around rectangular column (large eddy simulation (LES)), pressure distribution, streak line, time line, $Re = 22\ 000$ (page 261)

Plate 3 Turbulent velocity distribution in a clean room (finite element method); flow velocity from air ventilation system 38 cm/s (page 268)

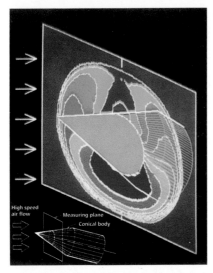

Plate 4 Density distribution of flow around a conical body flying at supersonic velocity, Mach number 2, angle of attack 20° (blue: low, red: high) (laser holographic interferometer + computer tomography) (page 287)

Plate 5 Density distribution on rotating fan blades and spinner, outer diameter 450 mm, speed of rotation 12 700 rpm (blue: low, red: high) (finite difference method) (page 289)

(a)

(b)

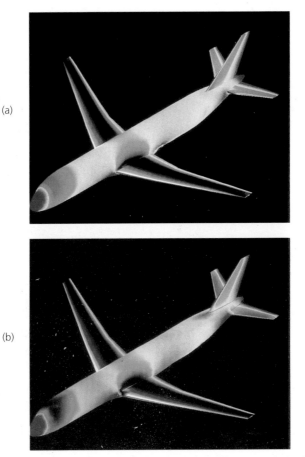

Plate 6 Pressure distribution on the surface of a transonic aircraft: (a) numerical simulation (boundary element method); (b) experimental result (measured pressure value); Mach number 0.6, angle of attack 0° (blue: low, red: high) (page 271)

Plate 7 Flow around an automobile (smoke method) in air, flow velocity 6 m/s, model 1:5, $Re = 2 \times 10^5$ (page 282)

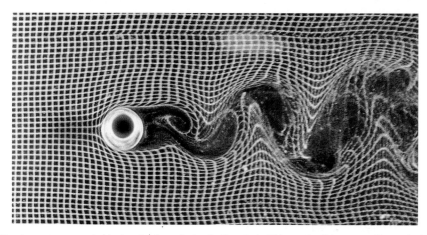

Plate 8 Kármán vortex street behind a circular cylinder (hydrogen bubble method) in water, flow velocity 2.6 m/s, diameter of cylinder 8 mm, $Re = 195$ (page 283)

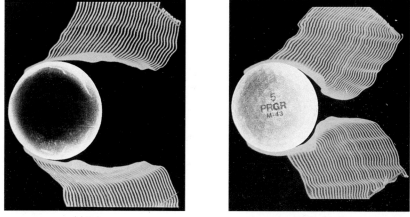

(a) Sphere   (b) Golf ball

Plate 9 Comparison of air flow around a golf ball and a sphere of equal size (spark tracing method) in air, flow velocity 23 m/s, diameter of ball 42.7 mm, $Re = 7 \times 10^4$ (page 283)

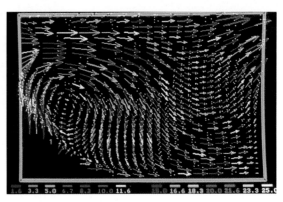

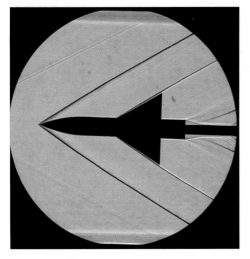

Plate 10 Supersonic flow around a simplified supersonic aircraft (AGARD-B model), (colour Schlieren method) (page 285)

Plate 11 Velocity vectors of flow over a circular cylinder (PTV) in water, flow velocity 1.2 m/s, diameter of cylinder 38.3 mm, *Re* = 3545, tracer: plastic particles of diameter 0.5 mm, velocity regions shown by colour (in mm/s) (page 286)

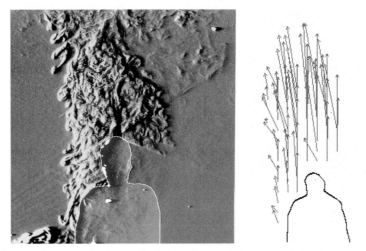

Plate 12 Natural convection around a human body (density correlation method). Maximum ascending velocity is about 0.2 m/s (page 286)

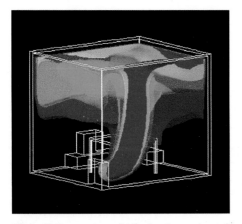

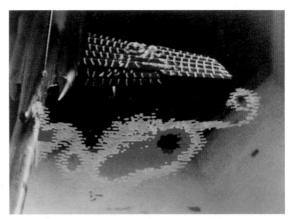

Plate 13 Temperature distribution in an air-conditioned room (isosurface manifestation method); red colour shows warmer areas (page 289)

Plate 14 Total head pattern behind horizontal tail (pressure sensors and light-emitting diodes combination method) (page 289)

Introduction to Fluid Mechanics

Introduction to Fluid Mechanics

Y. NAKAYAMA
Former Professor, Tokai University, Japan

UK Editor **R. F. BOUCHER**
Principal and Vice-Chancellor, UMIST, UK

BUTTERWORTH
HEINEMANN

OXFORD   AUCKLAND   BOSTON   JOHANNESBURG   MELBOURNE   NEW DELHI

Butterworth-Heinemann
Linacre House, Jordan Hill, Oxford OX2 8DP
225 Wildwood Avenue, Woburn, MA 01801-2041
A division of Reed Educational and Professional Publishing Ltd

℞ A member of the Reed Elsevier plc group

This book is translated from Ryutai-no-Rikigaku (in Japanese)
Published by
YOKENDO CO. LTD
5-30-15, Hongo, Bunkyo-ku, Tokyo
113-0033, Japan
© 1998 by Yasuki Nakayama

First published in English in Great Britain by Arnold 1999
Reprinted with revisions by Butterworth-Heinemann 2000

© Y. Nakayama and R. F. Boucher 1999

British Library Cataloguing in Publication Data
A catalogue record for this book is available from the British Library

Library of Congress Cataloguing in Publication Data
A catalogue record for this book is available from the Library of Congress

ISBN 0 340 67649 3

Commissioning Editor: Matthew Flynn
Production Editor: Liz Gooster
Production Controller: Sarah Kett
Cover design: Terry Griffiths

Typeset in 10/12 pt Times by AFS Image Setters Ltd, Glasgow
Printed and bound in Great Britain by MPG, Bodmin, Cornwall

Contents

About the authors

Professor Yasuki Nakayama graduated from Waseda University and received his doctorate in mechanical engineering from the same university. He joined the National Railway Research Institute and conducted many research investigations in the area of fluid mechanics. He then became a Professor of Tokai University, Japan, where he taught and researched fluid mechanics and visualisation. He later became President of the Future Technology Research Institute, Japan.

Professor Nakayama has received many distinctions and awards for his outstanding research. He has been a Visiting Professor of Southampton University, UK, President of The Visualisation Society of Japan, and Director of The Japan Society of Mechanical Engineers. He has published 10 books and more than 150 research papers.

Professor Robert Boucher FEng studied mechanical engineering in London and at Nottingham University. He has held posts in the electricity industry and at the Universities of Nottingham, Belfast, Sheffield and at UMIST, where he is Principal & Vice-Chancellor. His research interests, published in over 120 papers, include flow measurement and visualisation, fluid transients and network simulation, magnetic separation, industrial ventilation and oil drilling technology.

Preface

This book was written as a textbook or guidebook on fluid mechanics for students or junior engineers studying mechanical or civil engineering. The recent progress in the science of visualisation and computational fluid dynamics is astounding. In this book, effort has been made to introduce students/engineers to fluid mechanics by making explanations easy to understand, including recent information and comparing the theories with actual phenomena.

Fluid mechanics has hitherto been divided into 'hydraulics', dealing with the experimental side, and 'hydrodynamics', dealing with the theoretical side. In recent years, however, both have merged into an inseparable single science. A great deal was contributed by developments in the science of visualisation and by the progress in computational fluid dynamics using advances in computers. This book is written from this point of view.

The following features are included in the book:

1. Many illustrations, photographs and items of interest are presented for easy reading.
2. Portrait sketches of 17 selected pioneers who contributed to the development of fluid mechanics are inserted, together with brief descriptions of their achievements in the field.
3. Related major books and papers are presented in footnotes to facilitate advanced study.
4. Exercises appear at the ends of chapters to test understanding of the chapter topic.
5. Special emphasis is placed on flow visualisation and computational fluid dynamics by including 14 colour plates to assist understanding.

Books and papers by senior scholars throughout the world are referenced, with special acknowledgements to some of them. Among these, Professor R. F. Boucher, one of my oldest friends, assumed the role of editor of the English edition and made numerous revisions and additions by checking the book minutely during his busy time as Principal and Vice-Chancellor of UMIST. Another is Professor K. Kanayama of Musashino Academia Musicae who made many suggestions as my private language adviser. In addition, Mr Matthew Flynn and Dr Liz Gooster of Arnold took much trouble over the tedious editing work. I take this opportunity to offer my deepest appreciation to them all.

Yasuki Nakayama

List of symbols

A area
a area (relatively small), velocity of sound
B width of channel
b width, thickness
C coefficient of discharge
C_c coefficient of contraction
C_D drag coefficient
C_f frictional drag coefficient
C_L lift coefficient
C_M moment coefficient
C_v coefficient of velocity
c integration constant, coefficient of Pitot tube, flow velocity coefficient
c_p specific heat at constant pressure
c_v specific heat at constant volume
D diameter, drag
D_f friction drag
D_p pressure drag, form drag
d diameter
E specific energy
e internal energy
F force
F_r Froude number
f coefficient of friction
g gravitational acceleration
H head
h head, clearance, loss of head, depth, enthalpy
I geometrical moment of inertia
i slope
J moment of inertia
K bulk modulus
k interference factor
k_d cavitation number
L length, power, lift

l length, mixing length
M mass, Mach number
m mass flow rate, mass (relatively small), strength of doublet, hydraulic mean depth
n polytropic exponent
P total pressure
p pressure
p_0 stagnation pressure, total pressure, atmospheric pressure
p_s static pressure
p_t total pressure
p_∞ pressure unaffected by body, static pressure
Q volumetric flow rate
q discharge quantity per unit time, quantity of heat per unit mass
R gas constant
Re Reynolds number
r radius (at any position)
r_0 radius
s specific gravity, entropy, wetted perimeter
T tension, absolute temperature, torque, thrust, period
t time
U velocity unaffected by body
u velocity (x-direction), peripheral velocity
V volume
v specific volume, mean velocity, velocity (y-direction), absolute velocity
v_* friction velocity
W weight
w velocity (z-direction), relative velocity
$w(z)$ complex potential
α acceleration, angle, coefficient of discharge
β compressibility
Γ circulation, strength of vortex
γ specific weight
δ boundary layer thickness
δ^* displacement thickness
ζ vorticity
η efficiency
θ angle, momentum thickness
κ ratio of specific heat
λ friction coefficient of pipe μ
μ coefficient of viscosity, dynamic viscosity
v kinematic viscosity, angle
ρ velocity potential
τ shear stress
ϕ angle, velocity potential
ψ stream function
ω angular velocity

1

History of fluid mechanics

1.1 Fluid mechanics in everyday life

There is air around us, and there are rivers and seas near us. 'The flow of a river never ceases to go past, nevertheless it is not the same water as before. Bubbles floating along on the stagnant water now vanish and then develop but have never remained.' So stated Chohmei Kamo, the famous thirteenth-century essayist of Japan, in the prologue of *Hohjohki*, his collection of essays. In this way, the air and the water of rivers and seas are always moving. Such a movement of gas or liquid (collectively called 'fluid') is called the 'flow', and the study of this is 'fluid mechanics'.

While the flow of the air and the water of rivers and seas are flows of our concern, so also are the flows of water, sewage and gas in pipes, in irrigation canals, and around rockets, aircraft, express trains, automobiles and boats. And so too is the resistance which acts on such flows.

Throwing baseballs and hitting golf balls are all acts of flow. Furthermore, the movement of people on the platform of a railway station or at the intersection of streets can be regarded as forms of flow. In a wider sense, the movement of social phenomena, information or history could be regarded as a flow, too. In this way, we are in so close a relationship to flow that the 'fluid mechanics' which studies flow is really a very familiar thing to us.

1.2. The beginning of fluid mechanics

The science of flow has been classified into hydraulics, which developed from experimental studies, and hydrodynamics, which developed through theoretical studies. In recent years, however, both have merged into the single discipline called fluid mechanics.

Hydraulics developed as a purely empirical science with practical techniques beginning in prehistoric times. As our ancestors settled to engage in farming and their hamlets developed into villages, the continuous supply of a proper quantity of water and the transport of essential food and

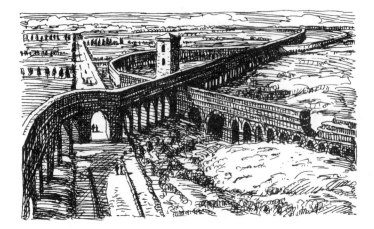

Fig. 1.1 Restored arch of Roman aqueduct in Compania Plain, Italy

materials posed the most important problems. In this sense, it is believed that hydraulics was born in the utilisation of water channels and ships.

Prehistoric relics of irrigation canals were discovered in Egypt and Mesopotamia, and it has been confirmed that canals had been constructed more than 4000 years BC.

Water in cities is said to have began in Jerusalem, where a reservoir to store water and a masonry channel to guide the water were constructed. Water canals were also constructed in Greece and other places. Above all, however, it was the Romans who constructed channels throughout the Roman Empire. Even today their remains are still visible in many places in Europe (Fig. 1.1).

The city water system in those days guided relatively clear water from far away to fountains, baths and public buildings. Citizens then fetched the water from water supply stations at high street corners etc. The quantity of water a day used by a citizen in those days is said to be approximately 180 litres. Today, the amount of water used per capita per day in an average household is said to be approximately 240 litres. Therefore, even about 2000 years ago, a considerably high level of cultural life occurred.

As stated above, the history of the city water system is very old. But in the development process of city water systems, in order to transport water effectively, the shape and size of the water conduit had to be designed and its

Fig. 1.2 Relief of ancient Egyptian ship

Fig. 1.3 Ancient Greek ship depicted on old vase

inclination or supply pressure had to be adjusted to overcome friction with the wall of the conduit. This gave rise to much invention and progress in overcoming hydraulic problems.

On the other hand, the origin of the ship is not clear, but it is easy to imagine the course of progress from log to raft, from manual propulsion to sails, and from river to ocean navigation. The Phoenicians and Egyptians built huge, excellent ships. The relief work shown in Fig. 1.2, which was made about 2700BC, clearly depicts a ship which existed at that time. The Greeks also left various records of ships. One of them is a beautiful picture of a ship depicted on an old Grecian vase, as shown in Fig. 1.3. As these objects indicate, it was by progress in shipbuilding and also navigation techniques that allowed much fundamental hydraulic knowledge to be accumulated.

Before proceeding to describe the development of hydraulics, the Renaissance period of Leonardo da Vinci in particular should be recalled. Popularly he is well known as a splendid artist, but he was an excellent scientist, too. He was so well versed in the laws of natural science that he stated that 'a body tries to drop down onto the earth through the shortest path', and also that 'a body gives air the same force as the resistance which air gives the body'. These statements preceded Newton's law of gravity and motion (law of action and reaction).

Particularly interesting in the history of hydraulics is Leonardo's note where a vast description is made of the movement of water, eddies, water waves, falling water, the destructive force of water, floating bodies, efflux and the flow in a tube/conduit to hydraulic machinery. As examples, Fig. 1.4 is a sketch of the flow around an obstacle, and Fig. 1.5 shows the development of vortices in the separation region. Leonardo was the first to find the least resistive 'streamline' shape.

Leonardo da Vinci (1452–1519)
An all-round genius born in Italy. His unceasing zeal for the truth and incomparable power of imagination are apparent in numerous sketches and astonishing design charts of implements, precise human anatomical charts and flow charts of fluids. He drew streamlines and vortices on these flow charts, which almost satisfy our present needs. It can therefore be said that he ingeniously suggested modern flow visualisation.

In addition, he made many discoveries and observations in the field of hydraulics. He forecast laws such as the drag and movement of a jet or falling water which only later scholars were to discover. Furthermore he advocated the observation of internal flow by floating particles in water, i.e. 'visualisation of the flow'. Indeed, Leonardo was a great pioneer who opened up the field of hydraulics. Excellent researchers followed in his footsteps, and hydraulics progressed greatly from the seventeenth to the twentieth century.

Fig. 1.4 Sketches from Leonardo da Vinci's notes (No. 1)

Fig. 1.5 Sketches from Leonardo da Vinci's notes (No. 2)

On the other hand, the advent of hydrodynamics, which tackles fluid movement both mathematically and theoretically, was considerably later than that of hydraulics. Its foundations were laid in the eighteenth century. Complete theoretical equations for the flow of non-viscous (non-frictional) fluid were derived by Euler (see page 59) and other researchers. Thereby, various flows were mathematically describable. Nevertheless, the computation according to these theories of the force acting on a body or the state of flow resulted in a very different outcome from the experimentally observed result.

In this way, hydrodynamics was thought to be without practical use. In the nineteenth century, however, it made such progress as to compete fully with hydraulics. One example of such progress was the derivation of the equation for the movement of a viscous fluid by Navier and Stokes. Unfortunately, since this equation has convection terms among the terms expressing the inertia (the terms expressing the force which varies from place to place), which renders the equation nonlinear, it was not an easy thing to obtain the analytical solution for general flows – only such special flows as laminar flow between parallel plates and in a round tube were solved.

Meanwhile, however, in 1869 an important paper was published which connected hydraulics and hydrodynamics. This was the report in which Kirchhoff, a German physicist (1824–87), computed the coefficient of contraction for the jet from a two-dimensional orifice as 0.611. This value coincided very closely with the experimental value for the case of an actual orifice – approximately 0.60.

As it was then possible to compute a value near the actual value, hydrodynamics was re-evaluated by hydraulics scholars. Furthermore, in the present age, with the progress in electronic computers and the development of various numerical techniques in hydrodynamics, it is now possible to obtain numerical solutions of the Navier–Stokes equation. Thus the barrier between hydraulics and hydrodynamics has now been completely removed, and the field is probably on the eve of a big leap into a new age.

2

Characteristics of a fluid

2.1 Fluid

Fluids are divided into liquids and gases. A liquid is hard to compress and as in the ancient saying 'Water takes the shape of the vessel containing it', it changes its shape according to the shape of its container with an upper free surface. Gas on the other hand is easy to compress, and fully expands to fill its container. There is thus no free surface.

Consequently, an important characteristic of a fluid from the viewpoint of fluid mechanics is its compressibility. Another characteristic is its viscosity. Whereas a solid shows its elasticity in tension, compression or shearing stress, a fluid does so only for compression. In other words, a fluid increases its pressure against compression, trying to retain its original volume. This characteristic is called compressibility. Furthermore, a fluid shows resistance whenever two layers slide over each other. This characteristic is called viscosity.

In general, liquids are called incompressible fluids and gases compressible fluids. Nevertheless, for liquids, compressibility must be taken into account whenever they are highly pressurised, and for gases compressibility may be disregarded whenever the change in pressure is small. Although a fluid is an aggregate of molecules in constant motion, the mean free path of these molecules is $0.06\,\mu m$ or so even for air of normal temperature and pressure, so a fluid is treated as a continuous isotropic substance.

Meanwhile, a non-existent, assumed fluid without either viscosity or compressibility is called an ideal fluid or perfect fluid. A fluid with compressibility but without viscosity is occasionally discriminated and called a perfect fluid, too. Furthermore, a gas subject to Boyle's–Charles' law is called a perfect or ideal gas.

2.2 Units and dimensions

All physical quantities are given by a few fundamental quantities or their combinations. The units of such fundamental quantities are called base units, combinations of them being called derived units. The system in which length, mass and time are adopted as the basic quantities, and from which the units of other quantities are derived, is called the absolute system of units.

2.2.1 Absolute system of units

MKS system of units
This is the system of units where the metre (m) is used for the unit of length, kilogram (kg) for the unit of mass, and second (s) for the unit of time as the base units.

CGS system of units
This is the system of units where the centimetre (cm) is used for length, gram (g) for mass, and second (s) for time as the base units.

International system of units (SI)
SI, the abbreviation of La Système International d'Unités, is the system developed from the MKS system of units. It is a consistent and reasonable system of units which makes it a rule to adopt only one unit for each of the various quantities used in such fields as science, education and industry.

There are seven fundamental SI units, namely: metre (m) for length, kilogram (kg) for mass, second (s) for time, ampere (A) for electric current, kelvin (K) for thermodynamic temperature, mole (mol) for mass quantity and candela (cd) for intensity of light. Derived units consist of these units.

Table 2.1 Dimensions and units

Quantity	Absolute system of units			
	α	β	γ	Units
Length	1	0	0	m
Mass	0	1	0	kg
Time	0	0	1	s
Velocity	1	0	-1	m/s
Acceleration	1	0	-2	m/s^2
Density	-3	1	0	kg/m^3
Force	1	1	-2	$N = kg\,m/s^2$
Pressure, stress	-1	1	-2	$Pa = N/m^2$
Energy, work	2	1	-2	J
Viscosity	-1	1	-1	Pa s
Kinematic viscosity	2	0	-1	m^2/s

2.2.2 Dimension

All physical quantities are expressed in combinations of base units. The index number of the combination of base units expressing a certain physical quantity is called the dimension, as follows.

In the absolute system of units the length, mass and time are respectively expressed by L, M and T. Put Q as a certain physical quantity and c as a proportional constant, and assume that they are expressed as follows:

$$Q = cL^{\alpha}M^{\beta}T^{\gamma} \quad \text{(SI)} \tag{2.1}$$

where α, β and γ are respectively called the dimensions of Q for L, M, T. Table 2.1 shows the dimensions of various quantities.

2.3 Density, specific gravity and specific volume

The mass per unit volume of material is called the density, which is generally expressed by the symbol ρ. The density of a gas changes according to the pressure, but that of a liquid may be considered unchangeable in general. The units of density are kg/m³ (SI). The density of water at 4°C and 1 atm (101 325 Pa, standard atmospheric pressure; see Section 3.1.1) is 1000 kg/m³.

The ratio of the density of a material ρ to the density of water ρ_w is called the specific gravity, which is expressed by the symbol s:

$$s = \rho/\rho_w \tag{2.2}$$

The reciprocal of density, i.e. the volume per unit mass, is called the specific volume, which is generally expressed by the symbol v:

$$v = 1/\rho \quad \text{(m}^3\text{/kg)} \tag{2.3}$$

Values for the density ρ of water and air under standard atmospheric pressure are given in Table 2.2.

Table 2.2 Density of water and air (standard atmospheric pressure)

Temperature (°C)	0	10	15	20	40	60	80	100
ρ (kg/m³) Water	999.8	999.7	999.1	998.2	992.2	983.2	971.8	958.4
Air	1.293	1.247	1.226	1.205	1.128	1.060	1.000	0.9464

2.4 Viscosity

As shown in Fig. 2.1, suppose that liquid fills the space between two parallel plates of area A each and gap h, the lower plate is fixed, and force F is needed to move the upper plate in parallel at velocity U. Whenever $Uh/v < 1500$ ($v = \mu/\rho$: kinematic viscosity), laminar flow (see Section 4.4) is maintained, and a linear velocity distribution, as shown in the figure, is obtained. Such a parallel flow of uniform velocity gradient is called the Couette flow.

In this case, the force per unit area necessary for moving the plate, i.e. the shearing stress (Pa), is proportional to U and inversely proportional to h. Using a proportional constant μ, it can be expressed as follows:

$$\tau = \frac{F}{A} = \mu\frac{U}{h} \tag{2.4}$$

The proportional constant μ is called the viscosity, the coefficient of viscosity or the dynamic viscosity.

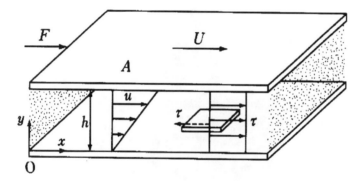

Fig. 2.1 Couette flow

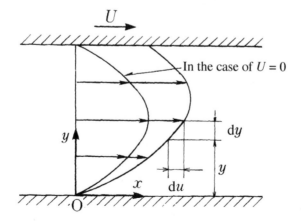

Fig. 2.2 Flow between parallel plates

Isaac Newton (1642–1727)
English mathematician, physicist and astronomer; studied at the University of Cambridge. His three big discoveries of the spectral analysis of light, universal gravitation and differential and integral calculus are only too well known. There are so many scientific terms named after Newton (Newton's rings and Newton's law of motion/viscosity/resistance) that he can be regarded as the greatest contributor to the establishment of modern natural science.

Newton's statue at Grantham near Woolsthorpe, his birthplace

Such a flow where the velocity u in the x direction changes in the y direction is called shear flow. Figure 2.1 shows the case where the fluid in the gap is not flowing. However, the velocity distribution in the case where the fluid is flowing is as shown in Fig. 2.2. Extending eqn (2.4) to such a flow, the shear stress τ on the section dy, distance y from the solid wall, is given by the following equation:

$$\tau = \mu \frac{du}{dy} \tag{2.5}$$

This relation was found by Newton through experiment, and is called Newton's law of viscosity.

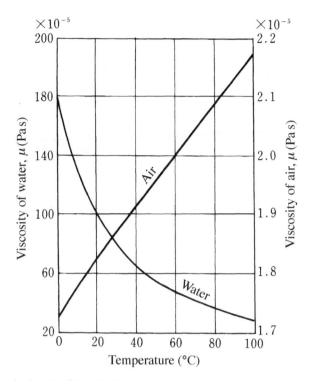

Fig. 2.3 Change in viscosity of air and of water under 1 atm

In the case of gases, increased temperature makes the molecular movement more vigorous and increases molecular mixing so that the viscosity increases. In the case of a liquid, as its temperature increases molecules separate from each other, decreasing the attraction between them, and so the viscosity decreases. The relation between the temperature and the viscosity is thus reversed for gas and for liquid. Figure 2.3 shows the change with temperature of the viscosity of air and of water.

The units of viscosity are Pa s (pascal second) in SI, and g/(cm s) in the CGS absolute system of units. 1g/(cm s) in the absolute system of units is called 1 P (poise) (since Poiseuille's law, stated in Section 6.3.2, is utilised for measuring the viscosity, the unit is named after him), while its 1/100th part is 1 cP (centipoise). Thus

$$1\,\mathrm{P} = 100\,\mathrm{cP} = 0.1\,\mathrm{Pa\,s}$$

The value v obtained by dividing viscosity μ by density ρ is called the kinematic viscosity or the coefficient of kinematic viscosity:

$$v = \frac{\mu}{\rho} \qquad (2.6)$$

Since the effect of viscosity on the movement of fluid is expressed by v, the name of kinematic viscosity is given. The unit is $\mathrm{m^2/s}$ regardless of the system of units. In the CGS system of units $1\,\mathrm{cm^2/s}$ is called 1 St (stokes) (since

Table 2.3 Viscosity and kinematic viscosity of water and air at standard atmospheric pressure

Temp. (°C)	Water		Air	
	Viscosity, μ (Pa s $\times 10^5$)	Kinematic viscosity, v (m²/s $\times 10^6$)	Viscosity, μ (Pa s $\times 10^5$)	Kinematic viscosity, v (m²/s $\times 10^6$)
0	179.2	1.792	1.724	13.33
10	130.7	1.307	1.773	14.21
20	100.2	1.004	1.822	15.12
30	79.7	0.801	1.869	16.04
40	65.3	0.658	1.915	16.98

Stokes' equation, to be stated in Section 9.3.3, is utilised for measuring the viscosity, it is named after him), while its 1/100th part is 1 cSt (centistokes). Thus

$$1\,St = 1 \times 10^{-4}\,m^2/s$$
$$1\,cSt = 1 \times 10^{-6}\,m^2/s$$

The viscosity μ and the kinematic viscosity v of water and air under standard atmospheric pressure are given in Table 2.3.

The kinematic viscosity v of oil is approximately 30–100 cSt. Viscosity

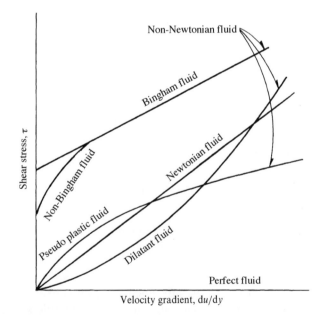

Fig. 2.4 Rheological diagram

sensitivity to temperature is expressed by the viscosity index VI,[1] a non-dimensional number. A VI of 100 is assigned to the least temperature sensitive oil and 0 to the most sensitive. With additives, the VI can exceed 100. While oil is used under high pressure in many cases, the viscosity of oil is apt to increase somewhat as the pressure increases.

For water, oil or air, the shearing stress τ is proportional to the velocity gradient du/dy. Such fluids are called Newtonian fluids. On the other hand, liquid which is not subject to Newton's law of viscosity, such as a liquid pulp, a high-molecular-weight solution or asphalt, is called a non-Newtonian fluid. These fluids are further classified as shown in Fig. 2.4 by the relationship between the shearing stress and the velocity gradient, i.e. a rheological diagram. Their mechanical behaviour is minutely treated by rheology, the science allied to the deformation and flow of a substance.

2.5 Surface tension

The surface of a liquid is apt to shrink, and its free surface is in such a state where each section pulls another as if an elastic film is being stretched. The tensile strength per unit length of assumed section on the free surface is called the surface tension. Surface tensions of various kinds of liquid are given in Table 2.4.

As shown in Fig. 2.5, a dewdrop appearing on a plant leaf is spherical in shape. This is also because of the tendency to shrink due to surface tension. Consequently its internal pressure is higher than its peripheral pressure. Putting d as the diameter of the liquid drop, T as the surface tension, and p as the increase in internal pressure, the following equation is obtained owing to the balance of forces as shown in Fig. 2.6:

$$\pi d T = \frac{\pi d^2}{4} \Delta p$$

or

$$\Delta p = 4T/d \qquad (2.7)$$

The same applies to the case of small bubbles in a liquid.

Table 2.4 Surface tension of liquid (20°C)

Liquid	Surface liquid	N/m
Water	Air	0.0728
Mercury	Air	0.476
Mercury	Water	0.373
Methyl alcohol	Air	0.023

[1] ISO 2909-1981

Fig. 2.5 A dewdrop on a taro leaf

Whenever a fine tube is pushed through the free surface of a liquid, the liquid rises up or falls in the tube as shown in Fig. 2.7 owing to the relation between the surface tension and the adhesive force between the liquid and the solid. This phenomenon is called capillarity. As shown in Fig. 2.8, d is the diameter of the tube, θ the contact angle of the liquid to the wall, ρ the density of liquid, and h the mean height of the liquid surface. The following equation is obtained owing to the balance between the adhesive force of liquid stuck to the wall, trying to pull the liquid up the tube by the surface tension, and the weight of liquid in the tube:

$$\pi d T \cos \theta = \frac{\pi d^2}{4} \rho g h$$

or

$$h = \frac{4T \cos \theta}{\rho g d} \tag{2.8}$$

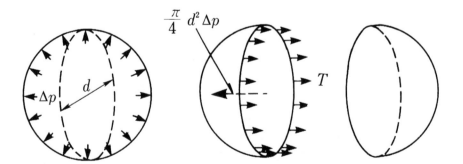

Fig. 2.6 Balance between the pressure increase within a liquid drop and the surface tension

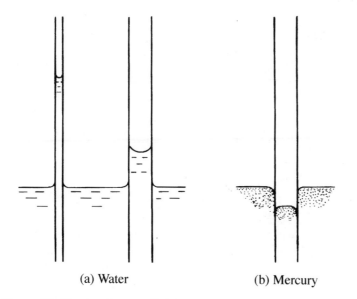

(a) Water　　　　　　　　(b) Mercury

Fig. 2.7 Change of liquid surface due to capillarity

Whenever water or alcohol is in direct contact with a glass tube in air under normal temperature, $\theta \simeq 0$. In the case of mercury, $\theta = 130°$–$150°$. In the case where a glass tube is placed in liquid,

$$\left.\begin{array}{ll} \text{for water} & h = 30/d \\ \text{for alcohol} & h = 11.6/d \\ \text{for mercury} & h = -10/d \end{array}\right\} \qquad (2.9)$$

(in mm). Whenever pressure is measured using a liquid column, it is necessary to pay attention to the capillarity correction.

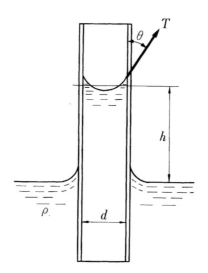

Fig. 2.8 Capillarity

2.6 Compressibility

As shown in Fig. 2.9, assume that fluid of volume V at pressure p decreased its volume by ΔV due to the further increase in pressure by Δp. In this case, since the cubic dilatation of the fluid is $\Delta V/V$, the bulk modulus K is expressed by the following equation:

$$K = \frac{\Delta p}{\Delta V/V} = -V\frac{dp}{dV} \tag{2.10}$$

Its reciprocal β

$$\beta = 1/K \tag{2.11}$$

is called the compressibility, whose value directly indicates how compressible the fluid is. For water of normal temperature/pressure $K = 2.06 \times 10^9$ Pa, and for air $K = 1.4 \times 10^5$ Pa assuming adiabatic change. In the case of water, $\beta = 4.85 \times 10^{-10}$ 1/Pa, and shrinks only by approximately 0.005% even if the atmospheric pressure is increased by 1 atm.

Putting ρ as the fluid density and M as the mass, since $\rho V = M = $ constant, assume an increase in density $\Delta\rho$ whenever the volume has decreased by ΔV, and

$$K = \rho\frac{\Delta p}{\Delta\rho} = \rho\frac{dp}{d\rho} \tag{2.12}$$

The bulk modulus K is closely related to the velocity a of a pressure wave propagating in a liquid, which is given by the following equation (see Section 13.2):

$$a = \sqrt{\frac{dp}{d\rho}} = \sqrt{\frac{K}{\rho}} \tag{2.13}$$

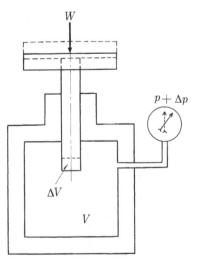

Fig. 2.9 Measuring of bulk modulus of fluid

Table 2.5 Gas constant R and ratio of specific heat κ

Gas	Symbol	Density (kg/m³) (0°C, 760 mm Hg)	R (SI) m²/(s² K)	$\kappa = c_p/c_v$
Helium	He	0.178 5	2 078.1	1.66
Air	–	1.293	287.1	1.402
Carbon monoxide	CO	1.250	296.9	1.400
Oxygen	O_2	0.089 9	4 124.8	1.409
Hydrogen	H_2	1.429	259.8	1.399
Carbon dioxide	CO_2	1.977	189.0	1.301
Methane	CH_4	0.717	518.7	1.319

2.7 Characteristics of a perfect gas

Let p be the pressure of a gas, v the specific volume, T the absolute temperature and R the gas constant. Then the following equation results from Boyle's–Charles' law:

$$pv = RT \tag{2.14}$$

This equation is called the equation of state of the gas, and $v = 1/\rho$ (SI) as shown in eqn (2.3). The value and unit of R varies as given in Table 2.5.

A gas subject to eqn (2.14) is called a perfect gas or an ideal gas. Strictly speaking, all real gases are not perfect gases. However, any gas at a considerably higher temperature than its liquefied temperature may be regarded as approximating to a perfect gas.

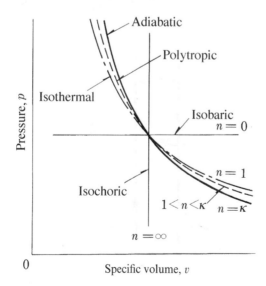

Fig. 2.10 State change of perfect gas

The change in state of a perfect gas is expressed by the following equation:

$$pv^n = \text{constant}$$

where n is called the polytropic exponent. As this value changes from 0 to ∞, as shown in Fig. 2.10, the state of gas makes five kinds of changes known as isobaric, isothermal, polytropic, adiabatic and isochoric changes. In particular, in the case of adiabatic change, $n = \kappa$ is obtained. Here κ is the ratio of specific heat at constant pressure c_p to specific heat at constant volume c_v, called the ratio of specific heats (isentropic index). Its value for various gases is given in Table 2.5.

2.8 Problems

1. Derive the SI unit of force from base units.

2. Express the viscosity and the kinematic viscosity in SI units.

3. The density of water at 4°C and 1 atm is 1000 kg/m^3. Obtain the specific volume v under such conditions.

4. Obtain the pressure in SI (Pa) necessary for shrinking the volume of water by 1% at normal temperature and pressure. Assume the compressibility of water $\beta = 4.85 \times 10^{-10} \, 1/\text{Pa}$.

5. When two plates are placed vertically on liquid as shown in Fig. 2.11, derive the equation showing the increased height of the liquid surface between the plates due to capillarity. Also when flat plates of glass are used with a 1 mm gap, what is the increased height of the water surface?

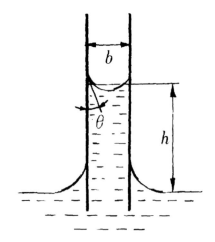

Fig. 2.11

6. Water at 20°C contains a bubble of diameter 1 mm. How much higher is the internal pressure of this bubble compared with the outside pressure?

7. How much force is necessary to lift a ring, diameter 20 mm, made of fine wire, and placed on the surface of water at 20°C?

8. As shown in Fig. 2.12, a cylinder of diameter 122 mm and length 200 mm is placed inside a concentric long pipe of diameter 125 mm. An oil film is introduced in the gap between the pipe and the cylinder. What force is necessary to move the cylinder at a velocity of 1 m/s? Assume that the dynamic viscosity of oil is 30 cSt and the specific gravity is 0.9.

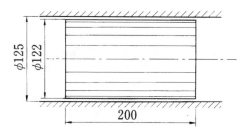

Fig. 2.12

9. Calculate the velocity of sound propagating in water at 20°C. Assume that the bulk modulus of water $K = 2.2 \times 10^9$ Pa.

3

Fluid statics

Fluid statics is concerned with the balance of forces which stabilise fluids at rest. In the case of a liquid, as the pressure largely changes according to its height, it is necessary to take its depth into account. Furthermore, even in the case of relative rest (e.g. the case where the fluid is stable relative to its vessel even when the vessel is rotating at high speed), the fluid can be regarded as being at rest if the fluid movement is observed in terms of coordinates fixed upon the vessel.

3.1 Pressure

When a uniform pressure acts on a flat plate of area A and a force P pushes the plate, then

$$p = P/A \tag{3.1}$$

In this case, p is the pressure and P is the pressure force. When the pressure is not uniform, the pressure acting on the minute area ΔA is expressed by the following equation:

$$p = \lim_{\Delta A \to 0} \frac{\Delta P}{\Delta A} = \frac{dP}{dA} \tag{3.2}$$

3.1.1 Units of pressure

The unit of pressure is the pascal (Pa), but it is also expressed in bars or metres of water column (mH_2O).[1] The conversion table of pressure units is given in Table 3.1. In addition, in some cases atmospheric pressure is used:

$$1\,\text{atm} = 760\,\text{mmHg (at 273.15 K}, g = 9.806\,65\,\text{m/s}^2) = 101\,325\,\text{Pa} \tag{3.3}$$

1 atm is standard 1 atmospheric pressure in meteorology and is called the standard atmospheric pressure.

[1] Refer to the spread of 'aqua' at the end of this chapter (p. 37).

Table 3.1 Conversion of pressure units

Name of unit	Unit	Conversion
Pascal	Pa	$1\,Pa = 1\,N/m^2$
Bar	bar	$1\,bar = 0.1\,MPa$
Water column metre	mH_2O	$1\,mH_2O = 9\,806.65\,Pa$
Atmospheric pressure	atm	$1\,atm = 101\,325\,Pa$
Mercury column metre	mHg	$1\,mHg = 1/0.76\,atm$
Torr	torr	$1\,torr = 1\,mm\,Hg$

3.1.2 Absolute pressure and gauge pressure

There are two methods used to express the pressure: one is based on the perfect vacuum and the other on the atmospheric pressure. The former is called the absolute pressure and the latter is called the gauge pressure. Then,

$$gauge\ pressure = absolute\ pressure - atmospheric\ pressure$$

In gauge pressure, a pressure under 1 atmospheric pressure is expressed as a negative pressure. This relation is shown in Fig. 3.1. Most gauges are constructed to indicate the gauge pressure.

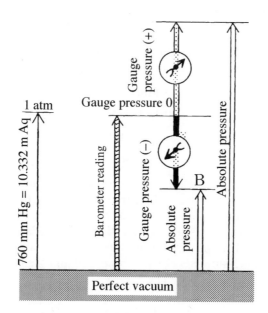

Fig. 3.1 Absolute pressure and gauge pressure

3.1.3 Characteristics of pressure

The pressure has the following three characteristics.

1. The pressure of a fluid always acts perpendicular to the wall in contact with the fluid.
2. The values of the pressure acting at any point in a fluid at rest are equal regardless of its direction. Imagine a minute triangular prism of unit width in a fluid at rest as shown in Fig. 3.2. Let the pressure acting on the small surfaces dA_1, dA_2 and dA be p_1, p_2 and p respectively. The following equations are obtained from the balance of forces in the horizontal and vertical directions:

$$p_1 dA_1 = pdA \sin \theta$$
$$p_2 dA_2 = pdA \cos \theta + \tfrac{1}{2} dA_1 dA_2 \rho g$$

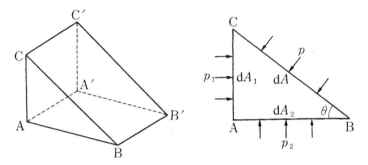

Fig. 3.2 Pressure acting on a minute triangular prism

The weight of the triangle pillar is doubly infinitesimal, so it is omitted. From geometry, the following equations are obtained:

$$dA \sin \theta = dA_1$$
$$dA \cos \theta = dA_2$$

Therefore, the following relation is obtained:

$$p_1 = p_2 = p \qquad (3.4)$$

Since angle θ can be given any value, values of the pressure acting at one point in a fluid at rest are equal regardless of its direction.

3. The fluid pressure applied to a fluid in a closed vessel is transmitted to all parts at the same pressure value as that applied (Pascal's law).

   In Fig. 3.3, when the small piston of area A_1 is acted upon by the force F_1, the liquid pressure $p = F_1/A_1$ is produced and the large piston is acted upon by the force $F_2 = pA_2$. Thus

$$F_2 = F_1 \frac{A_2}{A_1} \qquad (3.5)$$

Blaise Pascal (1623–62)
French mathematician, physicist and philosopher. He had the ability of a highly gifted scientist even in early life, invented an arithmetic computer at 19 years old and discovered the principle of fluid mechanics that carries his name. Many units had appeared as the units of pressure, but it was decided to use the pascal in SI units in memory of his achievements.

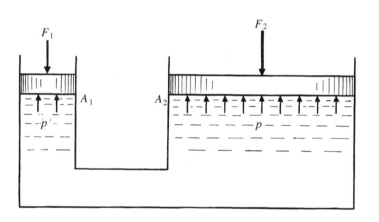

Fig. 3.3 Hydraulic press

So this device can create the large force F_2 from the small force F_1. This is the principle of the hydraulic press.

3.1.4 Pressure of fluid at rest

In general, in a fluid at rest the pressure varies according to the depth. Consider a minute column in the fluid as shown in Fig. 3.4. Assume that the sectional area is dA and the pressure acting upward on the bottom surface is p and the pressure acting downward on the upper surface (dz above the bottom surface) is $p + (dp/dz)dz$. Then, from the balance of forces acting on the column, the following equation is obtained:

$$p \, dA - \left(p + \frac{dp}{dz} dz \right) dA - \rho g \, dA \, dz = 0$$

or

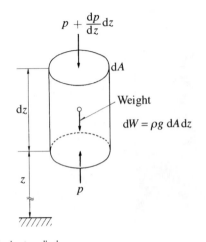

$$p + \frac{dp}{dz} dz$$

dA

Weight

$dW = \rho g\, dA\, dz$

dz

z

p

Fig. 3.4 Balance of vertical minute cylinder

$$\frac{dp}{dz} = -\rho g \qquad (3.6)$$

Since ρ is constant for liquid, the following equation ensues:

$$p = -\rho g \int dz = -\rho g z + c \qquad (3.7)$$

When the base point is set at z_0 below the upper surface of liquid as shown in Fig. 3.5, and p_0 is the pressure acting on that surface, then $p = p_0$ when $z = z_0$, so

$$c = p_0 + \rho g z$$

Substituting this equation into eqn (3.7),

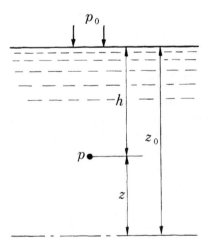

p_0

$-h$

z_0

p

z

Fig. 3.5 Pressure in liquid

$$p = p_0 + (z_0 - z)\rho g = p_0 + \rho g h \tag{3.8}$$

Thus it is found that the pressure inside a liquid increases in proportion to the depth.

For the case of a gas, let us study the relation between the pressure and the height of the atmosphere surrounding the earth. In this case, since the density of gas changes with pressure, it is not possible to integrate simply as in the case of a liquid. As the altitude increases, the temperature decreases. Assuming this temperature change to be polytropic, then $pv^n = $ constant is the defining relationship.

Putting the pressure and density at $z = 0$ (sea level) as p_0 and ρ_0 respectively, then

$$\frac{p}{\rho^n} = \frac{p_0}{\rho_0^n} \tag{3.9}$$

Substituting ρ into eqn (3.6),

$$dz = -\frac{dp}{\rho g} = -\frac{1}{g} \frac{p_0^{1/n}}{\rho_0} p^{-1/n} dp = -\frac{1}{g} \frac{p_0}{\rho_0} \left(\frac{p_0}{p}\right)^{1/n} d\left(\frac{p}{p_0}\right) \tag{3.10}$$

Integrating this equation from $z = 0$ (sea level),

$$z = \int_0^z dz = \frac{1}{g} \frac{n}{n-1} \frac{p_0}{\rho_0} \left[1 - \left(\frac{p}{p_0}\right)^{(n-1)/n} \right] \tag{3.11}$$

The relation between the height and the atmospheric pressure develops into the following equation by eqn (3.11):

$$\frac{p(z)}{p_0} = \left(1 - \frac{n-1}{n} \frac{\rho_0 g}{p_0} z \right)^{n/(n-1)} \tag{3.12}$$

Also, the density is obtained as follows from eqs (3.9) and (3.12):

$$\frac{\rho(z)}{\rho_0} = \left(1 - \frac{n-1}{n} \frac{\rho_0 g}{p_0} z \right)^{1/(n-1)} \tag{3.13}$$

When the absolute temperatures at sea level and at the point of height z are T_0 and T respectively, the following equation is obtained from eqn (2.14):

$$\frac{p}{\rho T} = \frac{p_0}{\rho_0 T_0} = R \tag{3.14}$$

From eqs (3.12)–(3.14)

$$\frac{T(z)}{T_0} = 1 - \frac{n-1}{n} \frac{\rho_0 g}{p_0} z \tag{3.15}$$

From eqn (3.15)

$$\frac{dT}{dz} = -\frac{n-1}{n} \frac{\rho_0 g}{p_0} T_0 = -\frac{n-1}{n} \frac{g}{R} \tag{3.16}$$

In aeronautics, it has been agreed to make the combined values of $p_0 = 101.325 \, \text{kPa}$, $T_0 = 288.15 \, \text{K}$ and $\rho_0 = 1.225 \, \text{kg/m}^3$ the standard atmospheric condition at sea level.[2] The temperature decreases by 0.65°C every 100 m of height in the troposphere up to approximately 1 km high, but is constant at -50.5°C from 1 km to 10 km high. For the troposphere, from the above values for p_0, T_0 and ρ_0 in eqn (3.10), $n = 1.235$ is obtained as the polytropic index.

3.1.5 Measurement of pressure

Manometer
A device which measures the fluid pressure by the height of a liquid column is called a manometer. For example, in the case of measuring the pressure of liquid flowing inside a pipe, the pressure p can be obtained by measuring the height of liquid H coming upwards into a manometer made to stand upright as shown in Fig. 3.6(a). When p_0 is the atmospheric pressure and ρ is the density, the following equation is obtained:

$$p = p_0 + \rho g H \tag{3.17}$$

When the pressure p is large, this is inconvenient because H is too high. So a U-tube manometer, as shown in Fig. 3.6(b), containing a high-density liquid such as mercury is used. In this case, when the density is ρ',

$$p + \rho g H = p_0 + \rho' g H'$$

or

$$p = p_0 + \rho' g H' - \rho g H \tag{3.18}$$

In the case of measuring the air pressure, $\rho' \gg \rho$, so $\rho g H$ in eqn (3.18) may be omitted. In the case of measuring the pressure difference between two pipes in both of which a liquid of density ρ flows, a differential manometer as

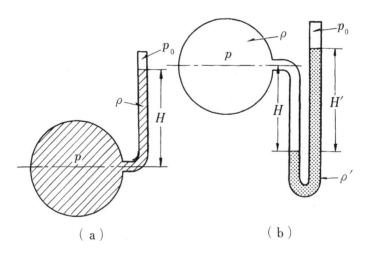

(a)   (b)

Fig. 3.6 Manometer

[2] ISO 2533-1975E.

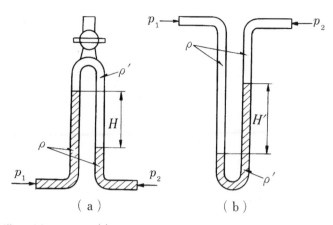

Fig. 3.7 Differential manometer (1)

shown in Fig. 3.7 is used. In the case of Fig. 3.7(a), where the differential pressure of the liquid is small, measurements are made by filling the upper section of the meter with a liquid whose density is less than that of the liquid to be measured, or with a gas. Thus

$$p_1 - p_2 = (\rho - \rho')gH \qquad (3.19)$$

and in the case where ρ' is a gas,

$$p_1 - p_2 = \rho g H \qquad (3.20)$$

Figure 3.7(b) shows the case when the differential pressure is large. This time, a liquid column of a larger density than the measuring fluid is used.

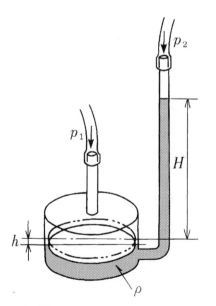

Fig. 3.8 Differential manometer (2)

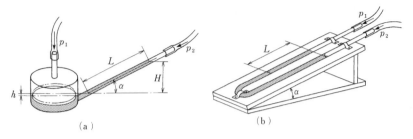

Fig. 3.9 Inclined manometer

Thus

$$p_1 - p_2 = (\rho' - \rho)gH' \tag{3.21}$$

and in the case where ρ is a gas,

$$p_1 - p_2 = \rho'gH' \tag{3.22}$$

A U-tube manometer as shown in Fig. 3.7 is inconvenient for measuring fluctuating pressure, because it is necessary to read both the right and left water levels simultaneously to measure the different pressure. For measuring the differential pressure, if the sectional area of one tube is made large enough, as shown in Fig. 3.8, the water column of height H could be measured by just reading the liquid surface level in the other tube because the surface fluctuation of liquid in the tank can be ignored.

To measure a minute pressure, a glass tube inclined at an appropriate angle as shown in Fig. 3.9 is used as an inclined manometer. When the angle of inclination is α and the movement of the liquid surface level is L, the differential pressure H is as shown in the following equation:

$$H = L\sin\alpha \tag{3.23}$$

Accordingly, if α is made smaller, the reading of the pressure is magnified. Besides this, Göttingen-type micromanometer, Chattock tilting micromanometer, etc., are used.

Elastic-type pressure gauge
An elastic-type pressure gauge is a type of pressure gauge which measures the pressure by balancing the pressure of the fluid with the force of deformation of an elastic solid. The Bourdon tube (invented by Eugene Bourdon, 1808–84) (Fig. 3.10), the diaphragm (Fig. 3.11), the bellows, etc., are widely employed for this type of pressure gauge.

Of these, the Bourdon tube pressure gauge (Bourdon gauge) of Fig. 3.10 is the most widely used in industry. A curved metallic tube of elliptical cross-section (Bourdon tube) is closed at one end which is free to move, but the other end is rigidly fixed to the frame. When the pressure enters from the fixed end, the cross-section tends to become circular so the free end moves outward. By amplifying this movement, the pressure values can be read. When the pressure becomes less than the atmospheric pressure (vacuum), the free end moves inward, so this gauge can be used as a vacuum gauge.

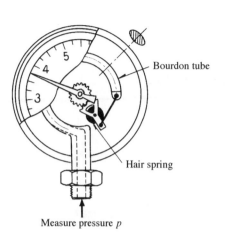

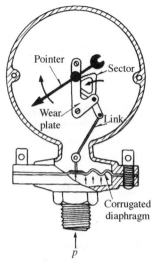

Fig. 3.10 Bourdon tube pressure gauge

Fig. 3.11 Diaphragm pressure gauge

Electric-type pressure gauge

The pressure is converted to the force or displacement passing through the diaphragm, Bourdon tube bellows, etc., and is detected as a change in an electrical property using a wire strain gauge, a semiconductor strain gauge

	PGM-C	PE-J, P
	(a) Small size	(b) For indicating pressure of an engine
Pressure range (MPa)	0.2–1.0	5–20
Natural frequency (kHz)	25–40	25–40
Response frequency (kHz)	5–8	5–8

Fig. 3.12 Wire strain gauge type of pressure transducer

(applied piezoresistance effect), etc. These types of pressure gauge are useful for measuring fluctuating pressures. Two examples of pressure gauges utilising the wire strain gauge are shown in Fig. 3.12.

3.2 Forces acting on the vessel of liquid

How large is the force acting on the whole face of a solid wall subject to water pressure, such as the bank of a dam, the sluice gate of a dam or the wall of a water tank? How large must the torque be to open the sluice gate of a dam? What is the force required to tear open a cylindrical vessel subject to inside pressure? Here, we will study forces like these.

3.2.1 Water pressure acting on a bank or a sluice gate

How large is the total force due to the water pressure acting on a bank built at an angle θ to the water surface (Fig. 3.13)? Here, disregarding the atmospheric pressure, the pressure acting on the surface is zero. The total pressure dP acting on a minute area dA is $\rho g h \, dA = \rho g y \sin \theta \, dA$. So, the total pressure P acting on the under water area of the bank wall A is:

$$P = \int_A dP = \rho g \sin \theta \int_A y \, dA$$

When the centroid[3] of A is G, its y coordinate is y_G and the depth to G is h_G, $\int_A y \, dA = y_G A$. So the following equation is obtained:

$$P = \rho g \sin \theta y_G A = \rho g h_G A \tag{3.24}$$

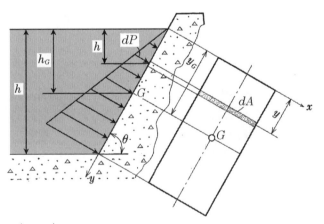

Fig. 3.13 Force acting on dam

---

[3] The centre of mass when the mass is distributed uniformly on the plane of some figure, namely the point applied to the centre of gravity, is called a centroid.

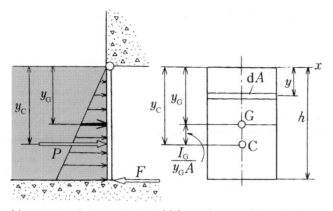

Fig. 3.14 Revolving power acting on water gate (1) (case where revolving axis of water gate is just on the water level)

So the total force P equals the product of the pressure at the centroid G and the underwater area of the bank wall.

Next, let us study a rectangular sluice gate as shown in Fig. 3.14. How large is the torque acting on its turning axis (the x axis)? The force P acting on the whole plane of the gate is $\rho g y_G A$ by eqn (3.24). The force acting on a minute area dA (a horizontal strip of the gate face) is $\rho g y \, dA$, the moment of this force around the x axis is $\rho g y \, dA \times y$ and the total moment on the gate is $\int \rho g y^2 \, dA = \rho g \int y^2 \, dA$. $\int y^2 \, dA$ is called the geometrical moment of inertia I_x for the x axis.

Now let us locate the action point of P (i.e. the centre of pressure C) at which a single force P produces a moment equal to the total sum of the moments around the turning axis (x axis) of the sluice gate produced by the total water pressure acting on all points of the gate. When the location of C is y_C,

$$P y_C = \rho g I_x \tag{3.25}$$

Now, when I_G is the geometrical moment of inertia of area for the axis which is parallel to the x axis and passes through the centroid G, the following relation exists:[4]

$$I_x = I_G + A y_G^2 \tag{3.26}$$

Values of I_G for a rectangular plate and for a circular plate are shown in Fig. 3.15.

Substitute eqn (3.26) into (3.25) to calculate y_C

$$y_C = y_G + \frac{I_G}{A y_G} = y_G + \frac{h^2}{12 y_G} \tag{3.27}$$

---

[4] *Parallel axis theorem*: The moment of inertia with respect to any axis equals the sum of the moment of inertia with respect to the axis parallel to this axis which passes through the centroid and the product of the sectional area and the square of the distance to the centroid from the former axis.

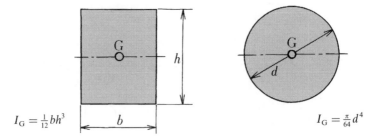

$$I_G = \tfrac{1}{12}bh^3 \qquad\qquad\qquad\qquad I_G = \tfrac{\pi}{64}d^4$$

Fig. 3.15 Geometrical moment of inertia for axis passing centroid G

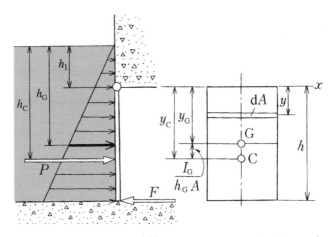

Fig. 3.16 Rotational force acting on water gate (2) (case where water gate is under water)

From eqn (3.27), it is clear that the action point C of the total pressure P is located deeper than the centroid G by $h^2/12y_G$.

The position of y_C in such a case where the sluice gate is located under the water surface as shown in Fig. 3.16 is given by eqn (3.28) where h_G is substituted for y_G in the second term on the right of eqn (3.27):

$$y_C = y_G + \frac{h^2}{12h_G} \tag{3.28}$$

3.2.2 Force to tear a cylinder

In the case of a thin cylinder where the inside pressure is acting outward, as shown in Fig. 3.17(a), what kind of force is required to tear this cylinder in the longitudinal direction? Now, consider the cylinder longitudinally half sectioned as shown in Fig. 3.17(b), with diameter d, length l and inside pressure p. The force acting on the assumed vertical centre wall ABCD is pdl which balances the force in the x direction acting outward on the cylinder wall. In other words, the force generated by the pressure in the x direction on a curved surface equals the pressure pdl, since the same pressure acts on the projected area of the curved surface. Furthermore, this force is the force $2Tl$

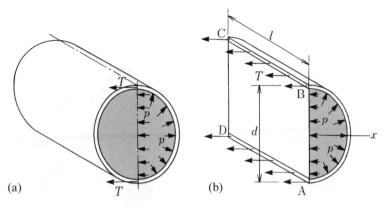

Fig. 3.17 Cylinder acted on by inertial pressure

(T is the force acting per unit length of wall which tears this cylinder longitudinally in halves along the lines BC and AD):

$$2Tl = pdl$$

or

$$T = pd/2 \qquad (3.29)$$

If the tensile stress due to T is lower than the allowable stress, safety is assured. By utilising this principle, a thin-walled pressure tank can be designed.

3.3 Why does a ship float?

Fluid pressure acts all over the wetted surface of a body floating in a fluid, and the resultant pressure acts in a vertical upward direction. This force is called buoyancy. The buoyancy of air is small compared with the gravitational force of the immersed body, so it is normally ignored.

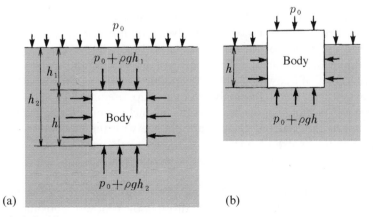

Fig. 3.18 Cube in liquid

Archimedes (287–212BC)
The greatest mathematician, physicist and engineer in ancient Greece, and the discoverer of the famous 'Principle of Archimedes'. Archimedes received guidance in astronomy from his father, an astronomer, and made astronomical observations since his early days. He invented a planetarium turned by hydropower and a screw pump. He carried out research in solid and fluid dynamics as well as on the lever, the centre of gravity and buoyancy. Archimedes was one of those scientists who are talented in both theory and practice.

Suppose that a cube is located in a liquid of density ρ as shown in Fig. 3.18. The pressure acting on the cube due to the liquid in the horizontal direction is balanced right and left. For the vertical direction, where the atmospheric pressure is p_0, the force F_1 acting on the upper surface A is expressed by the following equation:

$$F_1 = (p_0 + \rho g h_1)A \qquad (3.30)$$

The force F_2 acting on the lower surface is

$$F_2 = (p_0 + \rho g h_2)A \qquad (3.31)$$

So, when the volume of the body in the liquid is V, the resultant force F from the pressure acting on the whole surface of the body is

$$F = F_2 - F_1 = \rho g(h_2 - h_1)A = \rho g h A = \rho g V \qquad (3.32)$$

The same applies to the case where a cube is floating as shown in Fig. 3.18(b). From this equation, the body in the liquid experiences a buoyancy equal to the weight of the liquid displaced by the body. This result is known as Archimedes' principle. The centre of gravity of the displaced liquid is called 'centre of buoyancy' and is the point of action of the buoyancy force.

Next, let us study the stability of a ship. Figure 3.19 shows a ship of weight W floating in the water with an inclination of small angle θ. The location of the centroid G does not change with the inclination of the ship. But since the centre of buoyancy C moves to the new point C', a couple of forces $Ws = Fs$ is produced and this couple restores the ship's position to stability.

The forces of the couple Ws are called restoring forces. The intersecting point M on the vertical line passing through the centre of buoyancy C' (action line of the buoyancy F) and the centre line of the ship is called the

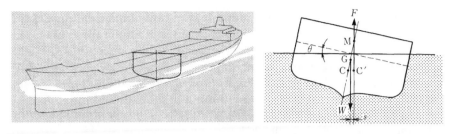

Fig. 3.19 Stability of a ship

metacentre, and GM is called the metacentric height.[5] As shown in the figure, if M is located higher than G, the restoring force acts to stabilise the ship, but if M is located lower than G, the couple of forces acts to increase the roll of the ship and so make the ship unstable.

3.4 Relatively stationary state

When a vessel containing a liquid moves in a straight line or rotates, if there is no relative flow of the liquid while the vessel and liquid move as a body, it is possible to treat this as the mechanics of a stationary state. This state is called a relatively stationary state.

3.4.1 Equiaccelerated straight-line motion

Suppose that a vessel filled with liquid is moving in a straight line at constant acceleration on the horizontal level as shown in Fig. 3.20. Further consider a minute element of mass m on the liquid surface, where its acceleration is α, the

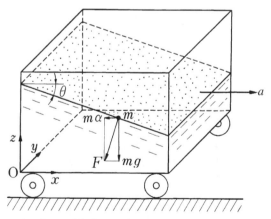

Fig. 3.20 Uniform accelerating straight-line motion

---

[5] How high is the metacentre of a real ship? It is said that the height of metacentre of a warship is about 0.8–1.2 m, a sailing ship 1.0–1.4 m and a large passenger ship 0.3–0.7 m. When these ships go out to sea the wave cycle is 12–13 seconds.

forces acting on m are gravity in a vertical downward direction $-mg$, and the inertial force in the reverse direction to the direction of acceleration $-m\alpha$.

There can be no force component normal to the direction of F, the resultant force of gravity and the inertial forces. Therefore, the pressure must be constant on the plane normal to the direction of F. In other words, this plane identifies the equipressure free surface.

When θ is the angle formed by the free surface and the x direction, the following relation is easily obtained:

$$\tan \theta = \alpha/g \tag{3.33}$$

If h is the depth measured in the vertical direction to the free surface, the acceleration in this direction is $\beta = F/m$. Therefore,

$$p = \rho\beta h \tag{3.34}$$

This is the same relation as the stationary state.

3.4.2 Rotational motion

Let us study the height of the water surface in the case where a cylindrical vessel filled with liquid is rotating at constant angular velocity ω. The movement at constant angular velocity like this is sometimes called gyrostatics, where the liquid surface poses a concave free surface. Then let us take cylindrical coordinates (r, θ, z) as shown in Fig. 3.21. Consider a minute element of mass m on the equipressure plane. The forces acting on it are $-mg$ due to the gravitational acceleration g in the vertical direction and $-mr\omega^2$ due to the centripetal acceleration $r\omega^2$ in the horizontal direction.

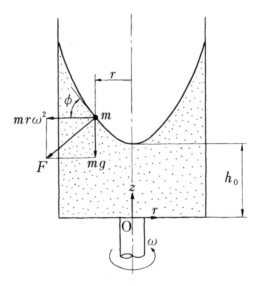

Fig. 3.21 Rotational motion around vertical axis

Since the vessel and liquid move in a body and the liquid stays in a relatively stationary state, the resultant force F is vertical to the free surface as in the previous case. If ϕ is the angle formed by the free surface and the horizontal direction,

$$\tan \phi = \frac{mr\omega^2}{mg} = \frac{r\omega^2}{g} \qquad (3.35)$$

but also

$$\tan \phi = \frac{dz}{dr}$$

Therefore,

$$\frac{dz}{dr} = \frac{r\omega^2}{g}$$

Putting c as a constant of integration,

$$z = \frac{\omega}{2g}r^2 + c \qquad (3.36)$$

If $z = h_0$ at $r = 0$, $c = h_0$ and the following equation is obtained from eqn (3.36):

$$z - h_0 = \frac{\omega^2 r^2}{2g} \qquad (3.37)$$

The free surface is now a rotating parabolic surface.

Spread of aqua

Aqua means water in Sanskrit, especially water offered to Buddha. The Italian word for water is *aqua*, and the Spanish is *agua*, both of which have the same etymological origin. Also, 'aqualung' is diving gear meaning water lung.

The Japanese word *aka* appeared in Japanese classics in the tenth, eleventh and thirteenth centuries. Furthermore, *aka* also means bilge water.

It is very interesting to know that Sanskrit *aqua* spread from India to Europe along the Silk Road and to Japan via China.

3.5 Problems

1. What is the water pressure on the sea bottom at a depth of 6500 m? The specific gravity of sea water is assumed to be 1.03.

2. Obtain the pressure p at point A in Figs 3.22(a), (b) and (c).

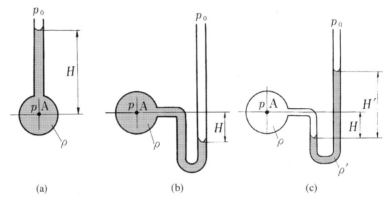

Fig. 3.22

(a)   (b)   (c)

3. Obtain the pressure difference $p_1 - p_2$ in Figs 3.23(a) and (b).

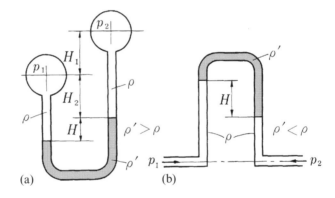

(a)   (b)

Fig. 3.23

4. On the inclined manometer in Fig. 3.24, whenever h changes by 1 mm, how high (in mm) is H? (Sectional area $A = 100a$ and $\alpha = 30°$.)

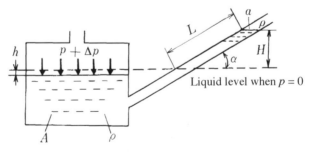

Liquid level when $p = 0$

Fig. 3.24

5. In the case shown in Fig. 3.25, an oblong board 3 m high and 5 m wide is placed vertically in water in such a manner that its upper face is 5 m deep. Obtain the force acting on this board and the location of the centre of pressure.

6. What are the respective forces F acting on the lower stays of the water gates in Figs 3.14 and 3.16, provided that the height b of the water gate is 3 m, the width is 1 m, and h_1 in Fig. 3.16 is 2 m?

7. A water gate 2 m high and 1 m wide is shown in Fig. 3.26. What is the force acting on the lower stay?

8. What is the force acting on a unit width of the dam wall shown in Fig. 3.27, if the water is 15 m deep and the wall is inclined at 60°? Furthermore, how far along the wall from the water surface is the action point of the force?

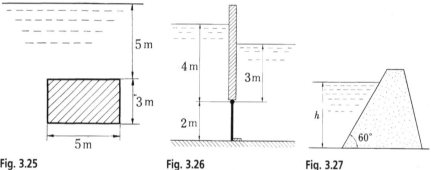

Fig. 3.25   Fig. 3.26   Fig. 3.27

9. As shown in Fig. 3.28, a circular water gate, diameter 2 m, is supported by a horizontal shaft. What is the moment around the shaft to keep the water gate closed?

10. A circular segment water gate, 5 m long, is set as shown in Fig. 3.29. Water is stored up to the upper face of the water gate. Obtain the magnitudes of the horizontal and vertical components of force and also the magnitude and the direction of the resultant force acting on this water gate.

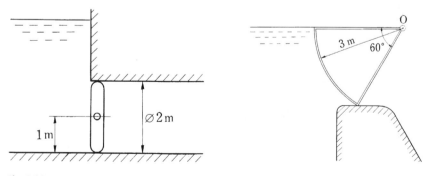

Fig. 3.28   Fig. 3.29

11. An iceberg of specific gravity 0.92 is floating on the sea with a specific gravity of 1.025. If the volume of the iceberg above the water level is $100\,\mathrm{m}^3$, what is the total volume of the iceberg?

12. As shown in Fig. 3.30 a body of specific gravity 0.8 is floating on the water. Obtain the height of its metacentre and the period of vibration whenever its side A is perturbed.[6] The effect of the additional mass of water can be omitted.

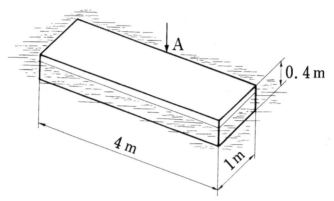

A

0. 4 m

4 m

1 m

Fig. 3.30

13. A cylindrical vessel of radius r_0 filled with water to height h is rotated around the central axis, and the difference in height of water level is h'. What is the rotational angular velocity? Furthermore, assuming $r_0 = 10\,\mathrm{cm}$ and $h = 18\,\mathrm{cm}$, obtain ω when $h' = 10\,\mathrm{cm}$ and also the number of revolutions per minute n when the cylinder bottom begins to appear.

---

[6] If h is the height of the metacentre, V the displacement volume, I the sectional secondary moment around the centre line of the water plane, and e the height from the centre of buoyancy to the centre of gravity, then

$$h = (I/V) - e$$

If J is the moment of inertia around the longitudinal axis passing through the centre of gravity, θ the inclination angle and m the mass, the movement equation for crosswise vibration is (whenever θ is small), then

$$J\frac{d^2\theta}{dt^2} = -mgh\theta$$

If T is the period, $k = \sqrt{J/m}$ the turning radius around the centre of gravity, then

$$T = 2\pi \frac{k}{\sqrt{gh}}$$

4

Fundamentals of flow

There are two methods for studying the movement of flow. One is a method which follows any arbitrary particle with its kaleidoscopic changes in velocity and acceleration. This is called the Lagrangian method. The other is a method by which, rather than following any particular fluid particle, changes in velocity and pressure are studied at fixed positions in space x, y, z and at time t. This method is called the Eulerian method. Nowadays the latter method is more common and effective in most cases.

Here we will explain the fundamental principles needed whenever fluid movements are studied.

4.1 Streamline and stream tube

A curve formed by the velocity vectors of each fluid particle at a certain time is called a streamline. In other words, the curve where the tangent at each point indicates the direction of fluid at that point is a streamline. Floating aluminium powder on the surface of flowing water and then taking a photograph, gives the flow trace of the powder as shown in Fig. 4.1(a). A streamline is obtained by drawing a curve following this flow trace. From the definition of a streamline, since the velocity vector has no normal component, there is no flow which crosses the streamline. Considering two-dimensional flow, since the gradient of the streamline is $\mathrm{d}y/\mathrm{d}x$, and putting the velocity in

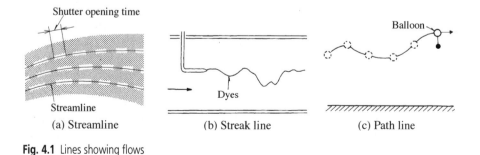

| (a) Streamline | (b) Streak line | (c) Path line |

Fig. 4.1 Lines showing flows

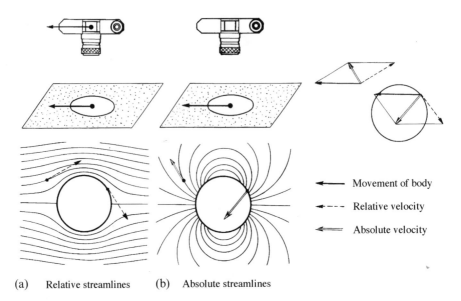

(a) Relative streamlines   (b) Absolute streamlines

Fig. 4.2 Relative streamlines and absolute streamlines

the x and y directions as u and v respectively, the following equation of the streamline is obtained:

$$dx/u = dy/v \qquad (4.1)$$

Whenever streamlines around a body are observed, they vary according to the relative relationship between the observer and the body. By moving both a cylinder and a camera placed in a water tank at the same time, it is possible to observe relative streamlines as shown in Fig. 4.2(a). On the other hand, by moving just the cylinder, absolute streamlines are observed (Fig. 4.2(b)).

In addition, the lines which show streams include the streak line and the path line. By the streak line is meant the line formed by a series of fluid particles which pass a certain point in the stream one after another. As shown in Fig. 4.1(b), by instantaneously catching the lines by injecting dye into the flow through the tip of a thin tube, the streak lines showing the turbulent flow can be observed. On the other hand, by the path line is meant the path of one particular particle starting from one particular point in the stream. As shown in Fig. 4.1(c), by recording on movie or video film a balloon released in the air, the path line can be observed.

In the case of steady flow, the above three kinds of lines all coincide.

By taking a given closed curve in a flow and drawing the streamlines passing all points on the curve, a tube can be formulated (Fig. 4.3). This tube is called a stream tube.

Since no fluid comes in or goes out through the stream tube wall, the fluid is regarded as being similar to a fluid flowing in a solid tube. This assumption is convenient for studying a fluid in steady motion.

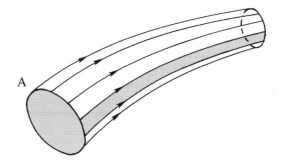

Fig. 4.3 Stream tube

4.2 Steady flow and unsteady flow

A flow whose flow state expressed by velocity, pressure, density, etc., at any position, does not change with time, is called a steady flow. On the other hand, a flow whose flow state does change with time is called an unsteady flow. Whenever water runs out of a tap while the handle is being turned, the flow is an unsteady flow. On the other hand, when water runs out while the handle is stationary, leaving the opening constant, the flow is steady.

4.3 Three-dimensional, two-dimensional and one-dimensional flow

All general flows such as a ball flying in the air and a flow around a moving automobile have velocity components in x, y and z directions. They are called three-dimensional flows. Expressing the velocity components in the x, y and z axial directions as u, v and w, then

$$u = u(x, y, z, t) \quad v = v(x, y, z, t) \quad w = w(x, y, z, t) \tag{4.2}$$

Consider water running between two parallel plates cross-cut vertically to the plates and parallel to the flow. If the flow states are the same on all planes parallel to the cut plane, the flow is called a two-dimensional flow since it can be described by two coordinates x and y. Expressing the velocity components in the x and y directions as u and v respectively, then

$$u = u(x, y, t) \qquad v = v(x, y, t) \tag{4.3}$$

and they can be handled more simply than in the case of three-dimensional flow.

As an even simpler case, considering water flowing in a tube in terms of average velocity, then the flow has a velocity component in the x direction only. A flow whose state is determined by one coordinate x only is called a one-dimensional flow, and its velocity u depends on coordinates x and t only:

$$u = u(x, t) \qquad\qquad (4.4)$$

In this case analysis is even simpler.

Although all natural phenomena are three dimensional, they can be studied as approximately two- or one-dimensional phenomena in many cases. Since the three-dimensional case has more variables than the two-dimensional case, it is not easy to solve the former. In this book three-dimensional formulae are omitted.

4.4 Laminar flow and turbulent flow

On a calm day with no wind, smoke ascending from a chimney looks like a single line as shown in Fig. 4.4(a). However, when the wind is strong, the smoke is disturbed and swirls as shown in Fig. 4.4(b) or diffuses into the peripheral air. One man who systematically studied such states of flow was Osborne Reynolds.

Reynolds used the device shown in Fig. 4.5. Coloured liquid was led to the entrance of a glass tube. As the valve was gradually opened by the handle, the coloured liquid flowed, as shown in Fig. 4.6(a), like a piece of thread without mixing with peripheral water.

When the flow velocity of water in the tube reached a certain value, he observed, as shown in Fig. 4.6(b) that the line of coloured liquid suddenly became turbulent on mingling with the peripheral water. He called the former flow the laminar flow, the latter flow the turbulent flow, and the flow velocity at the time when the laminar flow had turned to turbulent flow the critical velocity.

A familiar example is shown in Fig. 4.7. Here, whenever water is allowed to flow at a low velocity by opening the tap a little, the water flows out smoothly with its surface in the laminar state. But as the tap is gradually opened to let the water velocity increase, the flow becomes turbulent and opaque with a rough surface.

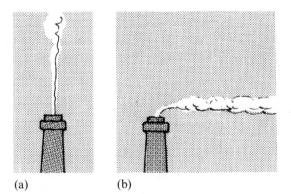

(a)   (b)

Fig. 4.4 Smoke from a chimney

Fig. 4.5 Reynolds' experiment[1]

(a) Laminar flow

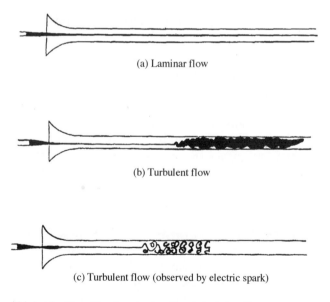

(b) Turbulent flow

(c) Turbulent flow (observed by electric spark)

Fig. 4.6 Reynolds' sketch of transition from laminar flow to turbulent flow

[1] Reynolds, O., *Philosophical Transactions of the Royal Society*, 174 (1883), 935.

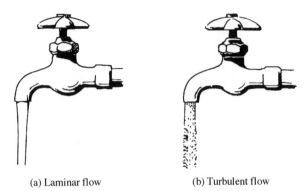

| (a) Laminar flow | (b) Turbulent flow |

Fig. 4.7 Water flowing from a faucet

4.5 Reynolds number

Reynolds conducted many experiments using glass tubes of 7, 9, 15 and 27 mm diameter and water temperatures from 4 to 44°C. He discovered that a laminar flow turns to a turbulent flow when the value of the non-dimensional quantity $\rho v d/\mu$ reaches a certain amount whatever the values of the average velocity v, glass tube diameter d, water density ρ and water viscosity μ. Later, to commemorate Reynolds' achievement,

$$Re = \frac{\rho v d}{\mu} = \frac{vd}{v} \quad (v \text{ is the kinematic viscosity}) \quad (4.5)$$

was called the Reynolds number. In particular, whenever the velocity is the critical velocity v_c, $Re_c = v_c d/v$ is called the critical Reynolds number. The value of Re_c is much affected by the turbulence existing in the fluid coming into the tube, but the Reynolds number at which the flow remains laminar, however agitated the tank water, is called the lower critical Reynolds number. This value is said to be 2320 by Schiller[2]. Whenever the experiment is made with calm tank water, Re_c turns out to have a large value, whose upper limit is called the higher critical Reynolds number. Ekman obtained a value of 5×10^4 for it .

4.6 Incompressible and compressible fluids

In general, liquid is called an incompressible fluid, and gas a compressible fluid. Nevertheless, even in the case of a liquid it becomes necessary to take compressibility into account whenever the liquid is highly pressurised, such

---

[2] Wien, W. und Harms, F., *Handbuch der Experimental Physik*, IV, 4 Teil, Akademische Verlagsgesellschaft (1932), 127.

Osborne Reynolds (1842–1912)
Mathematician and physicist of Manchester, England. His research covered all the fields of physics and engineering – mechanics, thermodynamics, electricity, navigation, rolling friction and steam engine performance. He was the first to clarify the phenomenon of cavitation and the accompanying noise. He discovered the difference between laminar and turbulent flows and the dimensionless number, the Reynolds number, which characterises these flows. His lasting contribution was the derivation of the momentum equation of viscous fluid for turbulent flow and the theory of oil-film lubrication.

as oil in a hydraulic machine. Similarly, even in the case of a gas, the compressibility may be disregarded whenever the change in pressure is small. As a criterion for this judgement, $\Delta \rho / \rho$ or the Mach number M (see Sections 10.4.1 and 13.3) is used, whose value, however, varies according to the nature of the situation.

4.7 Rotation and spinning of a liquid

Fluid particles running through a narrow channel flow, while undergoing deformation and rotation, are shown in Fig. 4.8.

Fig. 4.8 Deformation and rotation of fluid particles running through a narrowing channel

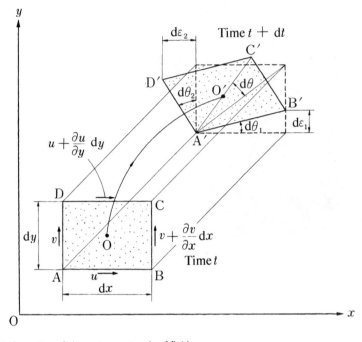

Fig. 4.9 Deformation of elementary rectangle of fluid

Now, assume that, as shown in Fig. 4.9, an elementary rectangle of fluid ABCD with sides dx, dy, which is located at O at time t moves to O′ while deforming itself to A′B′C′D′ time dt later.

AB in the x direction moves to A′B′ while rotating by dε_1, and AD in the y direction rotates by dε_2. Thus

$$d\varepsilon_1 = \frac{\partial v}{\partial x}\,dx\,dt \qquad d\varepsilon_2 = -\frac{\partial u}{\partial y}\,dy\,dt$$

$$d\theta_1 = \frac{d\varepsilon_1}{dx} = \frac{\partial v}{\partial x}\,dt \quad d\theta_2 = \frac{d\varepsilon_2}{dy} = -\frac{\partial u}{\partial y}\,dt$$

The angular velocities of AB and AD are ω_1 and ω_2 respectively:

$$\omega_1 = \frac{d\theta_1}{dt} = \frac{\partial v}{\partial x} \qquad \omega_2 = \frac{d\theta_2}{dt} = -\frac{\partial u}{\partial y}$$

For centre O, the average angular velocity ω is

$$\omega = \tfrac{1}{2}(\omega_1 + \omega_2) = \frac{1}{2}\left(\frac{\partial v}{\partial x} - \frac{\partial v}{\partial y}\right) \tag{4.6}$$

Putting the term in the large brackets of the above equation as

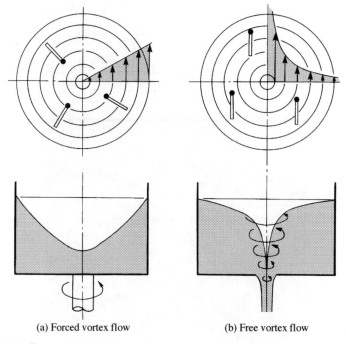

(a) Forced vortex flow   (b) Free vortex flow

Fig. 4.10 Vortex flow

$$\zeta = \frac{\partial v}{\partial x} - \frac{\partial u}{\partial y} \qquad (4.7)^3$$

gives what is called the vorticity for the z axis. The case where the vorticity is zero, namely the case where the fluid movement obeys

$$\frac{\partial v}{\partial x} - \frac{\partial u}{\partial y} = 0 \qquad (4.8)$$

is called irrotational flow.

As shown in Fig. 4.10(a), a cylindrical vessel containing liquid spins about the vertical axis at a certain angular velocity. The liquid makes a rotary

---

[3] In general, vector ς with the following components x, y, z for vector V (components x, y, z are u, v, w) is called the rotation or curl of vector V, which can be written as rot V, curl V and $\nabla \times V$ (∇ is called nabla). Thus

$$\zeta = \text{rot } V = \text{curl } V = \left[\frac{\partial w}{\partial y} - \frac{\partial v}{\partial z}, \frac{\partial u}{\partial z} - \frac{\partial w}{\partial x}, \frac{\partial v}{\partial x} - \frac{\partial u}{\partial y} \right]$$

Equation (4.7) is the case of two-dimensional flow where $w = 0$. ∇ is an operator which represents

$$i \frac{\partial}{\partial x} + j \frac{\partial}{\partial y} + k \frac{\partial}{\partial z}$$

where i, j, k are unit vectors on the x, y, z axes.

Fig. 4.11 Tornado

movement along the flow line, and, at the same time, the element itself rotates. This is shown in the upper diagram of Fig. 4.10(a), which shows how wood chips float, a well-studied phenomenon. In this case, it is a rotational flow, and it is called a forced vortex flow. Shown in Fig. 4.10(b) is the case of rotating flow which is observed whenever liquid is made to flow through a small hole in the bottom of a vessel. Although the liquid makes a rotary movement, its microelements always face the same direction without performing rotation. This case is a kind of irrotational flow called free vortex flow.

Hurricanes, eddying water currents and tornadoes (see Fig. 4.11) are familiar examples of natural vortices. Although the structure of these vortices is complex, the basic structure has a forced vortex at its centre and a free vortex on its periphery. Many natural vortices are generally of this type.

4.8 Circulation

As shown in Fig. 4.12, assuming a given closed curve s, the integrated v'_s (which is the velocity component in the tangential direction of the velocity v_s at a given point on this curve) along this same curve is called the circulation

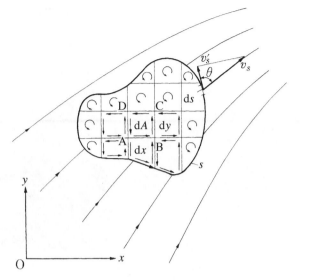

Fig. 4.12 Circulation

Γ. Here, counterclockwise rotation is taken to be positive. With the angle between v_s and v_s' as θ, then

$$\Gamma = \oint v_s' \, ds = \oint v_s \cos \theta \, ds \tag{4.9}$$

Next, divide the area surrounded by the closed curve s into microareas by lines parallel to the x and y axes, and study the circulation $d\Gamma$ of one such elementary rectangle ABCD (area dA), to obtain

$$d\Gamma = u \, dx + \left(v + \frac{\partial v}{\partial x} dx \right) dy - \left(u + \frac{\partial u}{\partial y} dy \right) dx - v \, dy = \left(\frac{\partial v}{\partial x} - \frac{\partial u}{\partial y} \right) dx \, dy$$

$$= \zeta \, dx \, dy = \zeta \, dA \tag{4.10}$$

ζ is two times the angular velocity ω of a rotational flow (eqn (4.6)), and the circulation is equal to the product of vorticity by area. Integrate eqn (4.10) for the total area, and the integration on each side cancels leaving only the integration on the closed curve s as the result. In other words,

$$\Gamma = \oint v_s' \, ds = \oint_A \zeta \, dA \tag{4.11}$$

From eqn (4.11) it is found that the surface integral of vorticity ζ is equal to the circulation. This relationship was introduced by Stokes, and is called Stokes' theorem. From this finding, whenever there is no vorticity inside a closed curve, then the circulation around it is zero. This theorem is utilised in fluid dynamics to study the flow inside the impeller of pumps and blowers as well as the flow around an aircraft wing.

George Gabriel Stokes (1819–1903)
Mathematician and physicist. He was born in Sligo in Ireland, received his education at Cambridge, became the professor of mathematics and remained in England for the rest of his life as a theoretical physicist. More than 100 of his papers were presented to the Royal Society, and ranged over many fields, including in particular that of hydrodynamics. His 1845 paper includes the derivation of the Navier–Stokes equations.

Reynolds' gleanings

Sir J. J. Thomson wrote:

> As I was taking the Engineering course, the Professor I had most to do with in my first three years at Owens was Professor Osborne Reynolds, the Professor of Engineering. He was one of the most original and independent of men and never did anything or expressed himself like anybody else. The result was that we had to trust mainly to Rankine's text books. Occasionally in the higher classes he would forget all about having to lecture and after waiting for ten minutes or so, we sent the janitor to tell him that the class was waiting. He would come rushing into the room pulling on his gown as he came through the door, take a volume of Rankine from the table, open it apparently at random, see some formula or other and say it was wrong. He then went up to the blackboard to prove this. He wrote on the board with his back to us, talking to himself, and every now and then rubbed it all out and said that was wrong. He would then start afresh on a new line, and so on. Generally, towards the end of lecture he would finish one which he did not rub out and say that this proved that Rankine was right after all.

Reynolds never blindly obeyed any scholar's view, even if he was an authority, without confirming it himself.

4.9 Problems

1. Put appropriate words in the blanks [   ] below.

   (a) A flow which does not change as time elapses is called a [   ]
   flow. [   ], [   ] and [   ] of flow in a steady flow are
   functions of position only, and most of the flows studied in
   hydrodynamics are steady flows. A flow which changes as time elapses
   is called an [   ] flow. [   ], [   ] and [   ] of
   flow in an unsteady flow are functions of [   ] and [   ].
   Flows such as when a valve is [   ]/[   ] or the [   ]
   from a tank belong to this flow.

   (b) The flow velocity is [   ] to the radius for a free vortex flow,
   and is [   ][   ] to the radius for a forced vortex flow.

2. When a cylindrical column of radius 5 cm is turned counterclockwise in
   fluid at 300 rpm, obtain the circulation of the fluid in contact with the
   column.

3. When water is running in a round tube of diameter 3 cm at a flow velocity
   of 2 m/s, is this flow laminar or turbulent? Assume that the kinematic
   viscosity of water is $1 \times 10^{-6}\,\text{m}^2/\text{s}$.

4. If the flow velocity is given by the following equations for a two-
   dimensional flow, obtain the equation of the streamline for this flow:

   $$u = kx \qquad v = -ky$$

5. If the flow velocities are given as follows, show respectively whether the
   flows are rotational or irrotational:

   (a) $u = -ky$ (b) $u = x^2 - y^2$ (c) $u = -\dfrac{ky}{x^2 + y^2}$

   $v = kx$ $v = -2xy$ $v = \dfrac{kx}{x^2 + y^2}$

   (k is constant).

6. Assuming that the critical Reynolds number of the flow in a circular pipe
   is 2320, obtain the critical velocity when water or air at 20°C is flowing in
   a pipe of diameter 1 cm.

7. A cylinder of diameter 1 m is turning counterclockwise at 500 rpm. Assuming that the fluid around the cylinder turns in contact with the column, obtain the circulation around it.

5

One-dimensional flow: mechanism for conservation of flow properties

General flows are three dimensional, but many of them may be studied as if they are one dimensional. For example, whenever a flow in a tube is considered, if it is studied in terms of mean velocity, it is a one-dimensional flow, which is studied very simply. Presented below are the methods of solution of those cases which may be studied as one-dimensional flows by using the continuity equation, energy equation and momentum equation.

5.1 Continuity equation

In steady flow, the mass flow per unit time passing through each section does not change, even if the pipe diameter changes. This is the law of conservation of mass.

For the pipe shown in Fig. 5.1 whose diameter decreases between sections 1 and 2, which have cross-sectional areas A_1 and A_2 respectively, and at which the mean velocities are v_1 and v_2 and the densities ρ_1 and ρ_2 respectively,

$$\rho_1 A_1 v_1 = \rho_2 A_2 v_2$$

namely,

$$\rho A v = \text{constant} \tag{5.1}$$

If the fluid is incompressible, e.g. water, with ρ being effectively constant, then

$$A v = \text{constant} \tag{5.2}$$

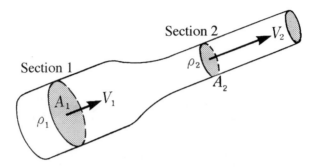

Fig. 5.1 Mass flow rate passing through any section is constant

ρAv is the mass of fluid passing through a section per unit time and this is called the mass flow rate. Av is that volume and this is called the volumetric flow rate, which is therefore constant is an incompressible pipe flow.

Equations (5.1) and (5.2) state that the flow is continuous, with no loss or gain, so these equations are called the continuity equations. They are an expression of the principle of conservation of mass when applied to fluid flow. It is clear from eqn (5.1) that the flow velocity is inversely proportional to the cross-sectional area of the pipe. When the diameter of the pipe is reduced, the flow velocity increases.

5.2 Conservation of energy

5.2.1 Bernoulli's equation

Consider a roller-coaster running with great excitement in an amusement park (Fig. 5.2). The speed of the roller-coaster decreases when it is at the top of the steep slope, and it increases towards the bottom. This is because the potential energy increases and kinetic energy decreases at the top, and the opposite occurs at the bottom. However, ignoring frictional losses, the sum of the two forms of energy is constant at any height. This is a manifestation of the principle of conservation of energy for a solid.

Figures 5.3(a) and (b) show the relationship between the potential energy of water (its level) and its kinetic energy (the speed at which it gushes out of the pipe).

A fluid can attain large kinetic energy when it is under pressure as shown in Fig. 5.3(c). A water hydraulic or oil hydraulic press machine is powered by the forces and energy due to such pressure.

In fluids, these three forms of energy are exchangeable and, again ignoring frictional losses, the total energy is constant. This is an expression of the law of conservation of energy applied to a fluid.

Fig. 5.2 Movement of roller-coaster

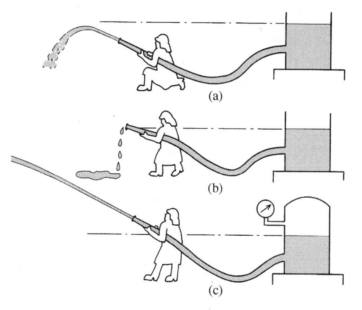

Fig. 5.3 Conservation of fluid energy

A streamline (a line which follows the direction of the fluid velocity) is chosen with the coordinates shown in Fig. 5.4. Around this line, a cylindrical element of fluid having the cross-sectional area dA and length ds is considered. Let p be the pressure acting on the lower face, and pressure $p + (\partial p/\partial s)ds$ acts on the upper face a distance ds away. The gravitational force acting on this element is its weight, $\rho g\, dA\, ds$. Applying Newton's second

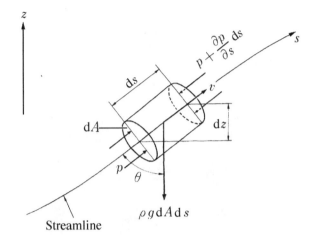

Fig. 5.4 Force acting on fluid on streamline

law of motion to this element, the resultant force acting on it, and producing acceleration along the streamline, is the force due to the pressure difference across the streamline and the component of any other external force (in this case only the gravitational force) along the streamline.

Therefore the following equation is obtained:

$$\rho \, dA \, ds \frac{dv}{dt} = -dA \frac{\partial p}{\partial s} ds - \rho g \, dA \, ds \cos \theta$$

or

$$\frac{dv}{dt} = -\frac{1}{\rho} \frac{\partial p}{\partial s} - g \cos \theta \tag{5.3}$$

The velocity may change with both position and time. In one-dimensional flow it therefore becomes a function of distance and time, $v = v(s, t)$. The change in velocity dv over time dt may be written as

$$dv = \frac{\partial v}{\partial t} dt + \frac{\partial v}{\partial s} ds$$

The acceleration is then

$$\frac{dv}{dt} = \frac{\partial v}{\partial t} + \frac{\partial v}{\partial t} \frac{ds}{dt} = \frac{\partial v}{\partial t} + v \frac{\partial v}{\partial s}$$

If the z axis is the vertical direction as shown in Fig. 5.4, then

$$\cos \theta = dz/ds$$

So eqn (5.3) becomes

$$\frac{\partial v}{\partial t} + v \frac{\partial v}{\partial s} = -\frac{1}{\rho} \frac{\partial p}{\partial s} - g \frac{dz}{ds} \tag{5.4}$$

In the steady state, $\partial v/\partial t = 0$ and eqn (5.4) would then become

Leonhard Euler (1707–83)
Mathematician born near Basle in Switzerland. A pupil of Johann Bernoulli and a close friend of Daniel Bernoulli. Contributed enormously to the mathematical development of Newtonian mechanics, while formulating the equations of motion of a perfect fluid and solid. Lost his sight in one eye and then both eyes, as a result of a disease, but still continued his research.

$$v\frac{dv}{ds} = -\frac{1}{\rho}\frac{dp}{ds} - g\frac{dz}{ds} \tag{5.5}$$

Equation (5.4) or (5.5) is called Euler's equation of motion for one-dimensional non-viscous fluid flow. In incompressible fluid flow with two unknowns (v and p), the continuity equation (5.2) must be solved simultaneously. In compressible flow, ρ becomes unknown, too. So by adding a third equation of state for a perfect gas (2.14), a solution can be obtained.

Equation (5.5) is integrated with respect to s to obtain a relationship between points a finite distance apart along the streamline. This gives

$$\frac{v^2}{2} + \int \frac{dp}{\rho} + gz = \text{constant} \tag{5.6}$$

and for an incompressible fluid ($\rho = \text{constant}$),

$$\frac{v^2}{2} + \frac{p}{\rho} + gz = \text{constant} \tag{5.7}$$

between arbitrary points, and therefore at all points, along a streamline.

Dividing each term in eqn (5.7) by g,

$$\frac{v^2}{2g} + \frac{p}{\rho g} + z = H = \text{constant} \tag{5.8}$$

Multiplying each term of eqn (5.7) by ρ,

$$\frac{\rho v^2}{2} + p + \rho gz = \text{constant} \tag{5.9}$$

The units of the terms in eqn (5.7) are m^2/s^2, which can be expressed as $kg\,m^2/(s^2\,kg\,)$. Since $kg\,m^2/s^2 = J$ (for energy), then $v^2/2$, p/ρ and gz in eqn

Daniel Bernoulli (1700–82)
Mathematician born in Groningen in the Netherlands. A good friend of Euler. Made efforts to popularise the law of fluid motion, while tackling various novel problems in fluid statics and dynamics. Originated the Latin word *hydrodynamica*, meaning fluid dynamics.

(5.7) represent the kinetic energy, energy due to pressure and potential energy respectively, per unit mass.

The terms of eqn (5.8) represent energy per unit weight, and they have the units of length (m) so they are commonly termed heads.

$$\frac{v^2}{2g} : \quad \text{velocity head}$$

$$\frac{p}{\rho g} : \quad \text{pressure head}$$

$$z : \quad \text{potential head}$$

$$H : \quad \text{total head}$$

The units of the terms of eqn (5.9) are $kg/(s^2\, m)$ expressing energy per unit volume. Thus, eqns (5.7) to (5.9) express the law of conservation of energy in that the sum of the kinetic energy, energy due to pressure and potential energy (i.e. the total energy) is always constant. This is Bernoulli's equation.

If the streamline is horizontal, then the term ρgh can be omitted giving the following:

$$\frac{\rho v^2}{2} + p_s = p_t \tag{5.10}$$

where $\rho v^2/2$ is called the dynamic pressure, p_s the static pressure, and p_t the total pressure or stagnation pressure.

Static pressure p_s can be detected, as shown in Fig. 5.5, by punching a small hole vertically in the solid wall face parallel to the flow.

As Bernoulli's theorem applies to a flow line, it is also applicable to the flow in a pipe line as shown in Fig. 5.6. Assume the pipe line is horizontal, and $z_1 = z_2$ in eqn (5.8). The following relative equation is obtained:

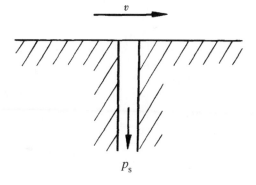

Fig. 5.5 Picking out of static pressure

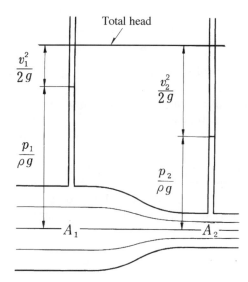

Fig. 5.6 Exchange between pressure head and velocity head

$$\frac{v_1^2}{2g} + \frac{p_1}{\rho g} = \frac{v_2^2}{2g} + \frac{p_2}{\rho g} \tag{5.11}$$

Also, from the continuity equation,

$$v_1 A_1 = v_2 A_2 \tag{5.12}$$

Consequently, whenever $A_1 > A_2$, then $v_1 < v_2$ and $p_1 > p_2$. In other words, where the flow channel is narrow (where the streamlines are dense), the flow velocity is large and the pressure head is low.

As shown in Fig. 5.7, whenever water flows from tank 1 to tank 2, the energy equations for sections 1, 2 and 3 are as follows from eqn (5.8):

$$\frac{v_1^2}{2} + \frac{p_1}{\rho} + z_1 = \frac{v_2^2}{2} + \frac{p_2}{\rho} + z_2 + h_2 = \frac{v_3^2}{2} + \frac{p_3}{\rho} + z_3 + h_3 \tag{5.13}$$

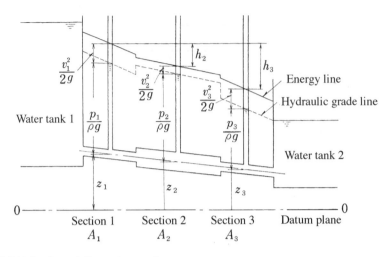

Fig. 5.7 Hydraulic grade line and energy line

h_2 and h_3 are the losses of head between section 1 and either of the respective sections.

In Fig. 5.7, the line connecting the height of the pressure heads at respective points of the pipe line is called the hydraulic grade line, while that connecting the heights of all the heads is called the energy line.

5.2.2 Application of Bernoulli's equation

Various problems on the one-dimensional flow of an ideal fluid can be solved by jointly using Bernoulli's theorem and the continuity equation.

Venturi tube

As shown in Fig. 5.8, a device where the flow rate in a pipe line is measured by narrowing a part of the tube is called a Venturi tube. In the narrowed part of the tube, the flow velocity increases. By measuring the resultant decreasing pressure, the flow rate in the pipe line can be measured.

Let A be the section area of the Venturi tube, v the velocity and p the pressure, and express the states of sections 1 and 2 by subscripts 1 and 2 respectively. Then from Bernoulli's equation

$$\frac{p_1}{\rho g} + \frac{v_1^2}{2g} + z_1 = \frac{p_2}{\rho g} + \frac{v_2^2}{2g} + z_2$$

Assuming that the pipe line is horizontal,

$$z_1 = z_2$$

$$\frac{v_2^2 - v_1^2}{2g} = \frac{p_1 - p_2}{\rho g}$$

From the continuity equation,

$$v_1 = v_2 A_2 / A_1$$

Giovanni Battista Venturi (1746–1822)
Italian physicist. After experiencing life as a priest, teacher and auditor, finally became a professor of experimental physics. Studied the effects of eddies and the flow rates at various forms of mouthpieces fitted to an orifice, and clarified the basic principles of the Venturi tube and the hydraulic jump in an open water channel.

Therefore,

$$v_2 = \frac{1}{\sqrt{1 - (A_2/A_1)^2}} \sqrt{2g\frac{p_1 - p_2}{\rho g}} \qquad (5.14)$$

and

$$\frac{p_1 - p_2}{\rho g} = H$$

Consequently, the flow rate

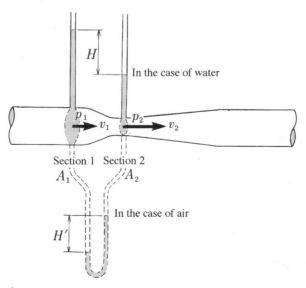

Fig. 5.8 Venturi tube

Henry de Pitot (1695–1771)
Born in Aramon in France. Studied mathematics and physics in Paris. As a civil engineer, undertook the drainage of marshy lands, construction of bridges and city water systems, and flood countermeasures. His books cover structures, land survey, astronomy, mathematics, sanitary equipment and theoretical ship steering in addition to hydraulics. The famous Pitot tube was announced in 1732 as a device to measure flow velocity.

$$Q = A_2 v_2 = \frac{A_2}{\sqrt{1 - (A_2/A_1)^2}} \sqrt{2gH} \qquad (5.15)$$

In the case where the flowing fluid is a gas, $p_1 - p_2$ is measured by a U-tube.

However, since there is some loss of energy between sections A_1 and A_2 in actual cases, the above equation is amended as follows:

$$Q = C \frac{A_2}{\sqrt{1 - (A_2/A_1)^2}} \sqrt{2gH} \qquad (5.16)$$

C is called the coefficient of discharge. It is determined through experiment. Equation (5.16) is also applicable to the case where the tube is inclined.

Pitot tube

Pitot, who was engaged in research work, hit upon an idea one day for a very simple measuring device of flow rate. It was a device where the lower end of a glass tube is bent by 90° and supported against the flow. The flow velocity was to be measured by measuring the increased height of the water level. It is said that, as soon as he had hit upon this idea, he rushed to the River Seine carrying a glass tube with a bent end. The result of an experiment as shown in Fig. 5.9 confirmed his expectation. The device incorporating that idea is shown in Fig. 5.10. This device is called a Pitot tube, and it is widely used even nowadays.

The tube is so designed that at the streamlined end a hole is opened in the face of the flow, while another hole in the direction vertical to the flow is used in order to pick out separate pressures.

Let p_A and v_A respectively be the static pressure and the velocity at position A of the undisturbed upstream flow. At opening B of the Pitot tube, the flow is stopped, making the velocity zero and the pressure p_B. B is called the stagnation point. Apply Bernoulli's equation between A and B,

Fig. 5.9 Pitot's first experiment

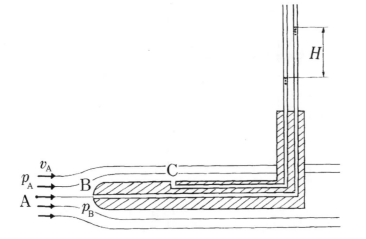

Fig. 5.10 Pitot tube

and

$$\frac{p_A}{\rho g} + \frac{v_A^2}{2g} = \frac{p_B}{\rho g}$$

or

$$v_A = \sqrt{2g\frac{p_B - p_C}{\rho}} \qquad (5.17)$$

In a parallel flow, the static pressure p_A is the same on the streamline adjacent to A and is detected by hole C normal to the flow. Thus, since $p_C = p_A$, eqn (5.17) becomes:

$$v_A = \sqrt{2\frac{p_B - p_C}{\rho}} \qquad (5.18)$$

And, since $(p_B - p_C)/\rho g = H$, the following equation is obtained:

$$v_A = \sqrt{2gH} \qquad (5.19)$$

In the case where the flowing fluid is a gas, $p_B - p_C$ is measured with a U-tube.

However, with an actual Pitot tube, since some loss occurs due to its shape and the fluid viscosity, the equation is modified as follows:

$$v_A = C_v\sqrt{2gH} \qquad (5.20)$$

where C_v is called the coefficient of velocity.

Flow through a small hole 1: *the case where water level does not change*
As shown in Fig. 5.11, we study here the case where water is discharging from a small hole on the side of a water tank. Such a hole is called an orifice. As

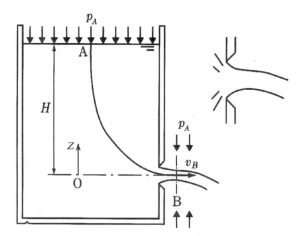

Fig. 5.11 Flow through a small hole (1)

shown in the figure, the spouting flow contracts to have its smallest section B a small distance from the hole. Here, it is conceived that the flow lines are almost parallel so that the pressures are uniform from the periphery to the centre of the flow. This part of the flow is called the vena contracta.

Assume that fluid particle A on the water surface has flowed down to section B. Then, from Bernoulli's theorem,

$$\frac{p_A}{\rho g} + \frac{v_A^2}{2g} + z_A = \frac{p_A}{\rho g} + \frac{v_B^2}{2g} + z_B$$

Assuming that the water tank is large and the water level does not change, at point A, $v_A = 0$ and $z_A = H$, while at point B, $z_B = 0$. If p_A is the atmospheric pressure, then

$$\frac{p_A}{\rho g} + H = \frac{p_A}{\rho g} + \frac{v_B^2}{2g}$$

or

$$v_B = \sqrt{2gH} \tag{5.21}$$

Equation (5.21) is called Torricelli's theorem.

Coefficient of contraction   Ratio C_c of area a_c of the smallest section of the discharging flow to area a of the small hole is called the coefficient of contraction, which is approximately 0.65:

$$a_c = C_c a \tag{5.22}$$

Coefficient of velocity   The velocity of spouting flow at the smallest section is less than the theoretical value $\sqrt{2gH}$ produced by the fluid velocity and the edge of the small hole. Ratio C_v of actual velocity v to $\sqrt{2gH}$ is called the coefficient of velocity, which is approximately 0.95:

$$v = C_v v_B = C_v \sqrt{2gH} \tag{5.23}$$

Coefficient of discharge   Consequently, the actual discharge rate Q is

$$Q = C_c a\, C_v v_B = C_c C_v a \sqrt{2gH} \tag{5.24}$$

Furthermore, setting $C_c C_v = C$, this can be expressed as follows:

$$Q = Ca\sqrt{2gH} \tag{5.25}$$

C is called the coefficient of discharge. For a small hole with a sharp edge, C is approximately 0.60.

Flow through a small hole 2: *the case where water level changes*
The theoretical flow velocity is

$$v = \sqrt{2gH}$$

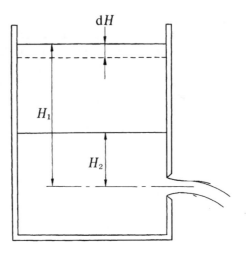

Fig. 5.12 Flow through a small hole (2)

Assume that dQ of water flows out in time dt with the water level falling by $-$dH (Fig. 5.12). Then

$$dQ = Ca\sqrt{2gH}\,dt = -dHA$$

$$dt = \frac{-A\,dH}{Ca\sqrt{2gH}}$$

$$\int_{t_1}^{t_2} dt = -\frac{A}{Ca\sqrt{2g}}\int_{H_1}^{H_2}\frac{dH}{\sqrt{H}}$$

The time needed for the water level to descend from H_1 to H_2 is

$$t_2 - t_1 = \frac{2A}{Ca\sqrt{2g}}(\sqrt{H_1} - \sqrt{H_2}) \qquad (5.26)$$

Flow through a small hole 3: *the section of water tank where the descending velocity of the water level is constant*
Assume that the bottom has a small hole of area a, through which water flows (Fig. 5.13), then

$$dQ = Ca\sqrt{2gH}\,dt = -dH\,A = -dH\,\pi r^2$$

Whenever the descending velocity of the water level ($-$d$H/$d$t = v$) is constant, the above equation becomes

$$v = -\frac{dH}{dt} = \frac{Ca\sqrt{2gH}}{\pi r^2} \qquad (5.27)$$

$$H = \left(\frac{\pi v}{Ca\sqrt{2g}}\right)^2 r^4 \qquad (5.28)$$

$$H \propto r^4 \qquad (5.29)$$

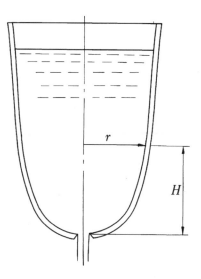

Fig. 5.13 Flow through a small hole (3)

In other words, whenever the section shape has a curve of r^4 against the vertical line, the descending velocity of the water level is constant.

Figure 5.14 shδws a water clock made in Egypt about 3400 years ago, which indicates the time by the position of the water level.

Fig. 5.14 Egyptian water clock 3400 years old (London Science Museum)

Weir

As shown in Fig. 5.15, in the case where a water channel is stemmed by a board or a wall, over which the water flows, such a board or wall is called a weir. A weir is used to adjust the flow rate.

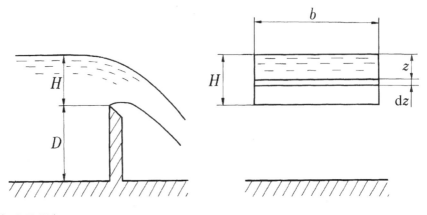

Fig. 5.15 Weir

In the figure, assume a minute depth dz at a given depth z from the water level. Let b be the width of the water channel and assume a minute area $b\,dz$ as an orifice. From Bernoulli's equation

$$v = \sqrt{2gz}$$

The flow rate dQ passing here is as follows assuming the coefficient of discharge is C:

$$dQ = Cb\,dz\sqrt{2gz}$$

Integrating the above equation,

$$Q = Cb\sqrt{2g}\int_0^H \sqrt{z}\,dz = \tfrac{2}{3}Cb\sqrt{2g}H^{3/2} \tag{5.30}$$

By measuring H, the discharge Q can be computed from eqn (5.30).

5.3 Conservation of momentum

5.3.1 Equation of momentum

A flying baseball can simply be caught with a glove. A moving automobile, however, is difficult to stop in a short time (Fig. 5.16). Therefore, the velocity is not sufficient to study the effects of bodily motion, but the product, Mv, of the mass M and the velocity v can be used as an indicator of the consequences of motion. This is called the linear momentum. By Newton's second law of motion, the change per unit time in the momentum of a body is equal to the force acting on the body.

Now, assume that a body of mass M (kg) will be at velocity v (m/s) in t seconds. The acting force F (N) is given by the following equation:

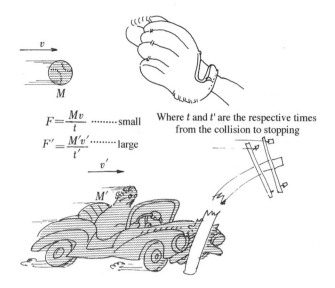

$$F = \frac{Mv}{t} \quad \text{········ small}$$

$$F' = \frac{M'v'}{t'} \quad \text{········ large}$$

Where t and t' are the respective times from the collision to stopping

Fig. 5.16 Car does not stop immediately

$$F = \frac{Mv_2 - Mv_1}{t} \tag{5.31}$$

In other words, the acting force is conserved as an increase in unit time in momentum. This is the law of conservation of momentum.

Whenever the reaction force of a jet or the force acting on a solid wall in contact with the flow is to be obtained, by using the change in momentum, such a force can be obtained comparatively simply without examining the complex internal phenomena.

In an actual computation, keeping in mind an assumed control volume in the flow, the relation between the change in momentum and the force within that volume is obtained by using the equation of momentum. In the case where fluid flows in a curved pipe as shown in Fig. 5.17, let ABCD be the control volume, A_1, A_2 the areas, v_1, v_2 the velocities, and p_1, p_2 the pressures of sections AB and CD respectively. Furthermore, let F be the force of fluid acting on the pipe; the force of the pipe acting on the fluid is $-F$. This force and the pressures acting on sections AB and CD act on the fluid, increasing the fluid momentum by such a combined force.[1] If F_x and F_y are the component forces in the x and y directions of F respectively, then from the equation of momentum,

$$\left. \begin{array}{l} -F_x + A_1 p_1 \cos \alpha_1 - A_2 p_2 \cos \alpha_2 = m(v_2 \cos \alpha_2 - v_1 \cos \alpha_1) \\ -F_y + A_1 p_1 \sin \alpha_1 - A_2 p_2 \sin \alpha_2 = m(v_2 \sin \alpha_2 - v_1 \sin \alpha_1) \end{array} \right\} \tag{5.32}$$

[1] Increase in momentum = momentum going out − momentum coming in.

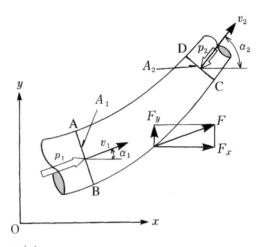

Fig. 5.17 Flow in a curved pipe

In this equation, m is the mass flow rate. If Q is the volumetric flow rate, then the following relation exists:

$$m = \rho Q = \rho A_1 v_1 = \rho A_2 v_2 = \rho Q$$

From eqn (5.32), F_x and F_y are given by

$$\left.\begin{aligned} F_x &= m(v_1 \cos \alpha_1 - v_2 \cos \alpha_2) + A_1 p_1 \cos \alpha_1 - A_2 p_2 \cos \alpha_2 \\ F_y &= m(v_1 \sin \alpha_1 - v_2 \sin \alpha_2) + A_1 p_1 \sin \alpha_1 - A_2 p_2 \sin \alpha_2 \end{aligned}\right\} \qquad (5.33)$$

Equation (5.32) is in the form where the change in momentum is equal to the force, but since m refers to mass per unit time, note that the equation shows that the time-sequenced change in momentum is equal to the force.

The combined force acting on the curved pipe can be obtained by the following equation:

$$F = \sqrt{F_x^2 + F_y^2} \qquad (5.34)$$

5.3.2 Application of equation of momentum

The equation of momentum is very effective when a fluid force acting on a body is studied.

Force of a jet

Let us study the case where, as shown in Fig. 5.18, a two-dimensional jet flow strikes an inclined flat plate at rest and breaks into upward and downward jets.

Assume that the internal pressure of the jet flow is equal to the external one and that no loss arises from the flow striking the flat plate. Since no loss occurs, it is assumed that the fluid flows out at the velocity v along the flat board after striking it. The control volume is conceived as shown in Fig. 5.18.

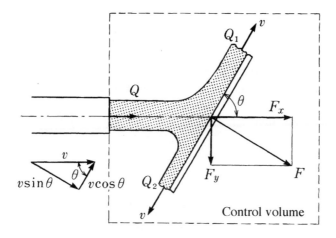

Fig. 5.18 Force of jet acting on a flat plate at rest

Examining the direction at right angles to the flat plate, since the velocity of the jet turns out to be zero after it has struck the flat board at $v \sin \theta$,

$$F = \rho Q v \sin \theta \qquad (5.35)$$

Force F_x acting in the direction of the jet is

$$F_x = F \sin \theta = \rho Q v \sin^2 \theta \qquad (5.36)$$

Force F_y acting in the direction at right angles to the jet is

$$F_y = F \cos \theta = \rho Q v \sin \theta \cos \theta \qquad (5.37)$$

Then the flow rate along the flat plate separates into Q_1 and Q_2. Let us obtain the change in the ratio of Q_1 to Q_2 according to the inclined angle θ. In this case, since no force acts along the flat board if the flow loss is disregarded, applying the equation of momentum to the direction along the flat board,

$$\rho Q v \cos \theta = \rho Q_1 v - \rho Q_2 v \quad Q \cos \theta = Q_1 - Q_2$$

Q_1 and Q_2 are obtained using the continuity equation $Q = Q_1 + Q_2$, and

$$Q_1 = Q(1 + \cos \theta)/2 \qquad (5.38)$$

$$Q_2 = Q(1 - \cos \theta)/2 \qquad (5.39)$$

In the case where the flat board in Fig. 5.18 moves in the same direction as the jet flow at velocity u, since the relative velocity of the jet flow compared with the flat board is $v - u$, the flow rate Q' reaching the flat board is given by

$$Q' = Q \frac{v - u}{v}$$

Since the change in velocity in the direction at right angles to the flat board is $(v - u) \sin \theta$, force F acting on the flat board is therefore

$$F = \rho Q'(v - u)\sin\theta = \rho Q \frac{(v - u)^2}{v}\sin\theta \tag{5.40}$$

Loss in a suddenly expanding pipe

For a suddenly expanding pipe as shown in Fig. 5.19, assume that the pipe is horizontal, disregard the frictional loss of the pipe, let h_s be the expansion loss, and set up an equation of energy between sections 1 and 2 as

$$\frac{p_1}{\rho g} + \frac{v_1^2}{2g} = \frac{p_2}{\rho g} + \frac{v_2^2}{2g} + h_s$$

or

$$h_s = \frac{p_1 - p_2}{\rho g} + \frac{v_1^2 - v_2^2}{2g} \tag{5.41}$$

Next, the streamlines in the smaller pipe are parallel at its very end, so the pressure there is p_1. And it can be considered that the pressure at the cross section is constant, so the pressure on the annular face at the pipe joint is also p_1. Apply the equation of momentum setting the control volume as shown in Fig. 5.19. Thus

$$\rho Q(v_2 - v_1) = (p_1 - p_2)A_2 \tag{5.42}$$

Since $Q = A_1 v_1 = A_2 v_2$, from the above equation,

$$\frac{p_1 - p_2}{\rho g} = \frac{Q}{A_2}\frac{v_2 - v_1}{g} = \frac{v_2}{g}(v_2 - v_1) \tag{5.43}$$

Substituting eqn (5.43) into (5.41),

$$h_s = \frac{(v_1 - v_2)^2}{2g} = \left(1 - \frac{A_1}{A_2}\right)^2 \frac{v_1^2}{2g} \tag{5.44}$$

is obtained. This h_s is called the Borda–Carnot head loss or simply the expansion loss.

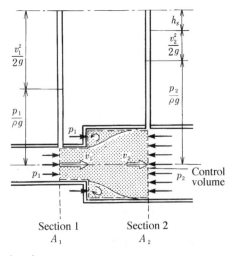

Fig. 5.19 Abruptly enlarging pipe

Jet pump

A jet pump is constructed as shown in Fig. 5.20. By making a water jet spout out into a larger water pipe, mixing with the surrounding water occurs so that it is carried out with that jet flow.

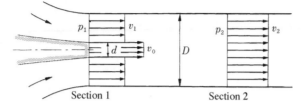

Fig. 5.20 Jet pump

If v_0 is the velocity of the jet discharging at section 1 and v_1 the velocity of the surrounding water, and assuming that mixing finishes at section 2 and the flow is then at uniform velocity v_2, then we have the following:

outflow momentum: $\dfrac{\pi D^2}{4}\rho v_2^2$

inflow in momentum: $\dfrac{\pi}{4}(D^2 - d^2)\rho v_1^2 + \dfrac{\pi}{4}d^2\rho v_0^2$

increase in momentum: $\dfrac{\pi}{4}\rho\big[D^2 v_2^2 - (D^2 - d^2)v_1^2 - d^2 v_0^2\big]$

force acting on the fluid: $\dfrac{\pi}{4}D^2(p_1 - p_2)$

By the law of momentum,

$$\rho\big[D^2 v_2^2 - (D^2 - d^2)v_1^2 - d^2 v_0^2\big] = D^2(p_1 - p_2)$$

Rearranging using the continuity equation,

$$p_2 - p_1 = \rho\frac{d^2}{D^2}\frac{D^2 - d^2}{D^2}(v_0 - v_1)^2 \tag{5.45}$$

This equation shows that $p_2 - p_1$ is always positive. In other words, a jet pump can force out water against the differential pressure.

Efficiency of a propeller

In the case shown in Fig. 5.21, a propeller of diameter D moving from right to left at velocity U can be considered as the case where a flow from left to right at velocity U strikes a propeller at rest. It can also be assumed that the fluid downstream has been accelerated to velocity $U + u$. Furthermore, the pressures upstream and downstream of the propeller are equally constant p.

From the changes in momentum and kinetic energy across the revolving face of the propeller, the thrust T is given by

$$T = \frac{\pi}{4}D^2\rho u\left(U + \frac{u}{2}\right) \tag{5.46}$$

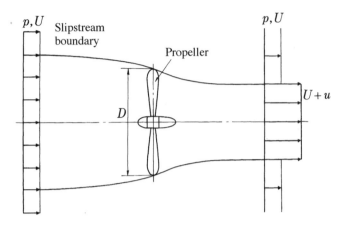

Fig. 5.21 Flows upstream and downstream of a propeller

and the efficiency η by

$$\eta = \frac{2}{2 + u/U} \tag{5.47}$$

Since the losses due to the fluid viscosity and the revolution of the wake are disregarded in this computation, this theory gives the attainable upper limit.

5.4 Conservation of angular momentum

5.4.1 Equation of angular momentum

The angular momentum in the case where a body of mass M is rotating at radius r and rotational velocity v is given by

Angular momentum = moment of inertia × angular velocity

$$= Mr^2 \times \frac{v}{r} = Mrv \tag{5.48}$$

The torque (rotational couple) on this body is given by

Torque = change of angular momentum

= moment of inertia × angular acceleration

$$\tag{5.49}$$

This is equivalent to Newton's second law of motion, and expresses the law of conservation of angular momentum.

Figure 5.22 shows a diagram of an ice skater. Whenever the skater revolves with the same angular momentum, if she spreads out her arms and stretches out one of her legs to enlarge the moment of inertia, she will slow down. This graphically expresses the relation of eqn (5.49).

If the relation of eqn (5.49) is applied to fluid flow, the torque acting on

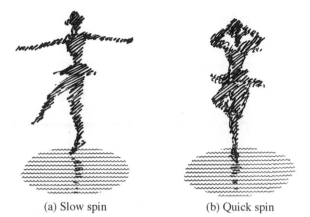

(a) Slow spin   (b) Quick spin

Fig. 5.22 Ice skater

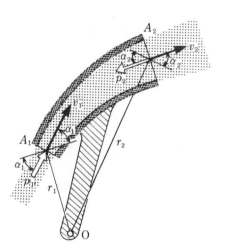

Fig. 5.23 Flow in curved tube supported so as to turn around shaft O

the shaft of a water wheel or a pump when the fluid runs over its rotating impeller can be obtained.

In the case where fluid is running in a curved tube as shown in Fig. 5.23, let T be the moment (torque),[2] which tries to turn the pipe around shaft O, generated by the force which the fluid between section A_1 and section A_2 exerts on the pipe wall. Then from the equation of angular momentum

$$T + A_2 p_2 r_2 \cos \alpha_2 - A_1 p_1 r_1 \cos \alpha_1 = m(r_2 v_2 \cos \alpha_2 - r_1 v_1 \cos \alpha_1) \qquad (5.50)$$

[2] The directions of rotation and torque are usually positive whenever they are counterclockwise.

5.4.2 Power of a water wheel or pump

Fluid flows at mass flow rate m along the blade in Fig. 5.24 due to rotation of the pump impeller. At radii r_1, r_2, the peripheral velocities are u_1, u_2 and v_1, v_2 are the absolute velocities at angles α_1, α_2 to them. The relative velocities to the impeller are w_1 and w_2. As seen from Fig. 5.24, since the direction of the pressures passes through the centre of the impeller, the second and third terms on the left eqn (5.50) turn out to be zero. The torque is as follows:

$$T = m(r_2 v_2 \cos \alpha_2 - r_1 v_1 \cos \alpha_1) \tag{5.51}$$

In this way, the torque acting on the impeller shaft can be obtained just from the states of the velocities at the inlet and outlet of the impeller.

If ω is the angular velocity of the impeller, the power L given to the shaft is

$$L = T\omega \tag{5.52}$$

The torque and power for a water wheel can be obtained similarly.

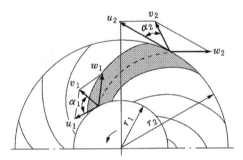

Fig. 5.24 Flow along blade of centrifugal pump

5.5 Problems

1. Derive Bernoulli's equation for steady flow by integrating Euler's equation of motion.

2. Find the flow velocities v_1, v_2 and v_3 in the conduit shown in Fig. 5.25. The flow rate Q is 800 L/min and the diameters d_1, d_2 and d_3 at sections 1, 2 and 3 are 50, 60 and 100 mm respectively.

3. Water is flowing in the conduit shown in Fig. 5.25. If the pressure p_1 at section 1 is 24.5 kPa, what are the pressures p_2 and p_3 at sections 2 and 3 respectively?

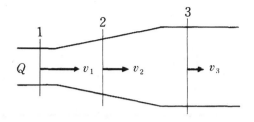

Fig. 5.25

4. In Fig. 5.26, air of flow rate Q flows into the centre through a pipe of radius r, and radially between two discs, and then flows out into the atmosphere. Obtain the pressure distribution between the discs. Also calculate the pressure force acting on the lower annular ring plate whose inner diameter is r_1 and outer diameter is r_2. Neglect frictional losses.

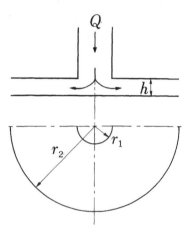

Fig. 5.26

5. In Fig. 5.26, if water flows at rate $Q = 0.013\,\text{m}^3/\text{s}$ radially between two discs of radius $r_2 = 30\,\text{cm}$ each from a pipe of radius $r_1 = 7\,\text{cm}$, obtain the pressure and the flow velocity at $r = 12\,\text{cm}$. Assume that $h = 0.3\,\text{cm}$ and neglect the frictional loss.

6. As shown in Fig. 5.27, a tank has a hole and $a \ll A$. Find the time necessary for the tank to empty.

7. As shown in Fig. 5.28, water flows out of a vessel through a small hole in the bottom. What is a suitable section shape to keep the velocity of descent of the water surface constant? Assume the volume of water in the vessel is $2\,l$, $R/d = 100$ (where R is the radius of the initial water surface in the vessel, d the small hole on the bottom), and the flow discharge coefficient of the small hole is $C = 0.6$. What should R and d be in order to manufacture a water clock for measuring 1 hour?

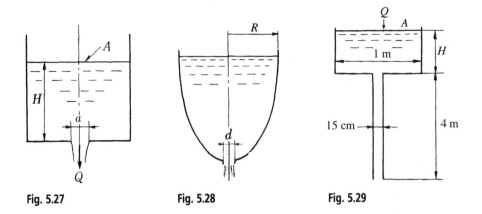

Fig. 5.27   Fig. 5.28   Fig. 5.29

8. In the case shown in Fig. 5.29, water at a flow rate of $Q = 0.2\,\mathrm{m^3/s}$ is supplied to the cylindrical water tank of diameter 1 m discharging through a round pipe of length 4 m and diameter 15 cm. How deep will the water in the tank be?

9. As shown in Fig. 5.30, a jet of water of flow rate Q and diameter d strikes the stationary plate at angle θ. Calculate the force on this stationary plate and its direction. Furthermore, if $\theta = 60°$, $d = 25\,\mathrm{mm}$ and $Q = 0.12\,\mathrm{m^3/s}$, obtain Q_1, Q_2 and F.

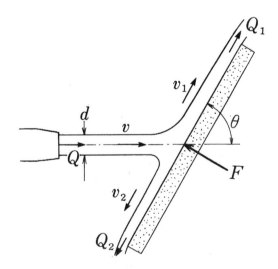

Fig. 5.30

10. As shown in Fig. 5.31, if water flows out of the tank of head 50 cm through the throttle, obtain the pressure at the throat.

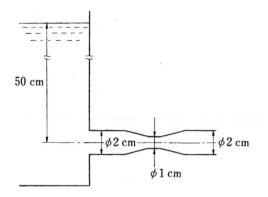

Fig. 5.31

11. Figure 5.32 shows a garden sprinkler. If the sprinkler nozzle diameter is 5 mm and the sprinkler velocity is 5 m/s, what is the rate of rotation? What torque is required to hold the sprinkler stationary? Assume there is no friction.

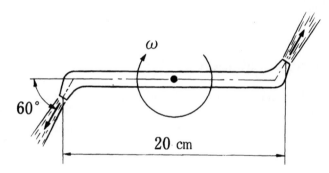

Fig. 5.32

12. A jet-propelled boat as shown in Fig. 5.33 is moving at a velocity of 10 m/s. The river is flowing against the boat at 5 m/s. Assuming the jet flow rate is 0.15 m³/s and its discharge velocity is 20 m/s, what is the propelling power of this boat? (Jet boats like this are actually in use.)

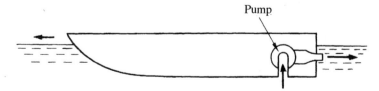

Fig. 5.33

6

Flow of viscous fluid

All fluids are viscous. In the case where the viscous effect is minimal, the flow can be treated as an ideal fluid, otherwise the fluid must be treated as a viscous fluid. For example, it is necessary to treat a fluid as a viscous fluid in order to analyse the pressure loss due to a flow, the drag acting on a body in a flow and the phenomenon where flow separates from a body. In this chapter, such fundamental matters are explained to obtain analytically the relation between the velocity, pressure, etc., in the flow of a two-dimensional incompressible viscous fluid.

6.1 Continuity equation

Consider the elementary rectangle of fluid of side dx, side dy and thickness b as shown in Fig. 6.1 (b being measured perpendicularly to the paper). The velocities in the x and y directions are u and v respectively. For the x

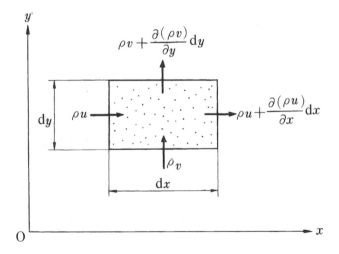

Fig. 6.1 Flow balance in a fluid element

direction, by deducting the outlet mass flow rate from the inlet mass flow rate, the fluid mass stored in the fluid element per unit time can be obtained, i.e.

$$\rho u b \, dy - \left[\rho u + \frac{\partial(\rho u)}{\partial x} \, dx \right] b \, dy = - \frac{\partial(\rho u)}{\partial x} b \, dx \, dy$$

Similarly, the fluid mass stored in it per unit time in the y direction is

$$-\frac{\partial(\rho v)}{\partial y} b \, dx \, dy$$

The mass of fluid element ($\rho b \, dx \, dy$) ought to increase by $\partial(\rho b \, dx \, dy/\partial t)$ in unit time by virtue of this stored fluid. Therefore, the following equation is obtained:

$$-\frac{\partial(\rho u)}{\partial x} b \, dx \, dy - \frac{\partial(\rho v)}{\partial y} b \, dx \, dy = \frac{\partial(\rho b \, dx \, dy)}{\partial t}$$

or

$$\frac{\partial \rho}{\partial t} + \frac{\partial(\rho u)}{\partial x} + \frac{\partial(\rho v)}{\partial y} = 0 \qquad (6.1)^1$$

Equation (6.1) is called the continuity equation. This equation is applicable to the unsteady flow of a compressible fluid. In the case of steady flow, the first term becomes zero.

For an incompressible fluid, ρ is constant, so the following equation is obtained:

$$\frac{\partial u}{\partial x} + \frac{\partial v}{\partial y} = 0 \qquad (6.2)^1$$

This equation is applicable to both steady and unsteady flows.

In the case of an axially symmetric flow as shown in Fig. 6.2, eqn (6.2) becomes, using cylindrical coordinates,

$$\frac{\partial u}{\partial x} + \frac{1}{r} \frac{\partial(rv)}{\partial r} = 0 \qquad (6.3)$$

As the continuity equation is independent of whether the fluid is viscous or not, the same equation is applicable also to an ideal fluid.

6.2 Navier–Stokes equation

Consider an elementary rectangle of fluid of side dx, side dy and thickness b as shown in Fig. 6.3, and apply Newton's second law of motion. Where the

---

[1] $\partial u/\partial x + \partial v/\partial y + \partial w/\partial z$ is generally called the divergence of vector V (whose components x, y, z are u, v, w) and is expressed as div V or ∇V. If we use this, eqns (6.1) and (6.2) (two-dimensional flow, so $w = 0$) are expressed respectively as the following equations:

$$\frac{\partial \rho}{\partial t} + \text{div}(\rho V) = 0 \quad \text{or} \quad \frac{\partial \rho}{\partial t} + \nabla(\rho V) = 0 \qquad (6.1)'$$

$$\text{div}(\rho V) = 0 \quad \text{or} \quad \nabla(\rho V) = 0 \qquad (6.2)'$$

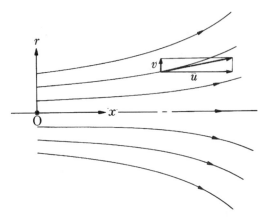

Fig. 6.2 Axially symmetric flow

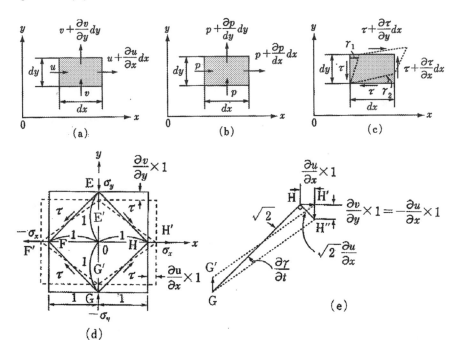

Fig. 6.3 Balance of forces on a fluid element: (a) velocity; (b) pressure; (c) angular deformation; (d) relation between tensile stress and shearing stress by elongation transformation of x direction; (e) velocity of angular deformation by elongation and contraction

forces acting on this element are $F(F_x, F_y)$, the following equations are obtained for the x and y axes respectively:

$$\left. \begin{array}{l} \rho b\, \mathrm{d}x\, \mathrm{d}y \dfrac{\mathrm{d}u}{\mathrm{d}t} = F_x \\[2mm] \rho b\, \mathrm{d}x\, \mathrm{d}y \dfrac{\mathrm{d}v}{\mathrm{d}t} = F_y \end{array} \right\}$$ (6.4)

The left-hand side of eqn (6.4) expresses the inertial force which is the product of the mass and acceleration of the fluid element. The change in velocity of this element is brought about both by the movement of position and by the progress of time. So the velocity change du at time dt is expressed by the following equation:

$$du = \frac{\partial u}{\partial t}dt + \frac{\partial u}{\partial x}dx + \frac{\partial u}{\partial y}dy$$

Therefore,

$$\frac{du}{dt} = \frac{\partial u}{\partial t} + \frac{\partial u}{\partial x}\frac{dx}{dt} + \frac{\partial u}{\partial y}\frac{dy}{dt} = \frac{\partial u}{\partial t} + u\frac{\partial u}{\partial x} + v\frac{\partial u}{\partial y}$$

Substituting this into eqn (6.4),

$$\left.\begin{aligned}\rho\left(\frac{\partial u}{\partial t} + u\frac{\partial u}{\partial x} + v\frac{\partial u}{\partial y}\right)b\,dx\,dy = F_x\\[2mm]\rho\left(\frac{\partial v}{\partial t} + u\frac{\partial v}{\partial x} + v\frac{\partial v}{\partial y}\right)b\,dx\,dy = F_y\end{aligned}\right\}\qquad(6.5)$$

Next, the force F acting on the elements comprises the body force $F_B(B_x, B_y)$, pressure force $F_p(P_x, P_y)$ and viscous force $F_s(S_x, S_y)$. In other words, F_x and F_y are expressed by the following equation:

$$\left.\begin{aligned}F_x = B_x + P_x + S_x\\F_y = B_y + P_y + S_y\end{aligned}\right\}\qquad(6.6)$$

Body force $F_b(B_x, B_y)$

(These forces act directly throughout the mass, such as the gravitational force, the centrifugal force, the electromagnetic force, etc.) Putting X and Y as the x and y axis components of such body forces acting on the mass of fluid, then

$$\left.\begin{aligned}B_x = X\rho b\,dx\,dy\\B_y = Y\rho b\,dx\,dy\end{aligned}\right\}\qquad(6.7)$$

For the gravitational force, $X = 0$, $Y = -g$.

Pressure force $F_p(P_x, P_y)$

Here,

$$\left.\begin{aligned}P_x = pb\,dy - \left(p + \frac{\partial p}{\partial x}dx\right)b\,dy = -\frac{\partial p}{\partial x}b\,dx\,dy\\[2mm]P_y = -\frac{\partial p}{\partial y}b\,dx\,dy\end{aligned}\right\}\qquad(6.8)$$

Viscous force $F_s(S_x, S_y)$

Force in the x direction due to angular deformation, S_{x1} Putting the strain of the small element of fluid $\gamma = \gamma_1 + \gamma_2$, the corresponding stress is expressed as $\tau = \mu \, \partial \gamma / \partial t$:

$$\tau = \mu \frac{\partial \gamma}{\partial t} = \mu \left(\frac{\partial \gamma_1}{\partial t} + \frac{\partial \gamma_2}{\partial t} \right) = \mu \left(\frac{\partial u}{\partial y} + \frac{\partial v}{\partial x} \right)$$

So,

$$S_{x1} = \frac{\partial \tau}{\partial y} b \, dx \, dy = \mu \left(\frac{\partial^2 u}{\partial y^2} + \frac{\partial^2 v}{\partial x \, \partial y} \right) b \, dx \, dy = \mu \left(\frac{\partial^2 u}{\partial y^2} - \frac{\partial^2 u}{\partial x^2} \right) b \, dx \, dy \quad (6.9)$$

Force in the x direction due to elongation transformation, S_{x2}   Consider the rhombus EFGH inscribed in a cubic fluid element ABCD of unit thickness as shown in Fig. 6.3(d), which shows that an elongated flow to x direction is a contracted flow to y direction. This deformation in the x and y directions produces a simple angular deformation seen in the rotation of the faces of the rhombus.

Now, calculating the deformation per unit time, the velocity of angular deformation $\partial \gamma / \partial t$ becomes as seen from Fig. 6.3(e).

$$\frac{\partial \gamma}{\partial t} = \frac{\sqrt{2} \frac{\partial u}{\partial x}}{\sqrt{2}} = \frac{\partial u}{\partial x}$$

Therefore, a shearing stress τ acts on the four faces of the rhombus EFGH.

$$\tau = \mu \frac{\partial \gamma}{\partial t} = \mu \frac{\partial u}{\partial x}$$

For equilibrium of the force on face EG due to the tensile stress σ_x and the shear forces on EH and HG due to τ

$$\sigma_x = 2 \times \sqrt{2} \tau \cos 45° = 2\tau$$

$$\sigma_x = 2\mu \frac{\partial u}{\partial x}$$

Considering the fluid element having sides dx, dy and thickness b, the tensile stress in the x direction on the face at distance dx becomes $\sigma_x + \frac{\partial \sigma_x}{\partial x} dx$. This stress acts on the face of area $b \, dy$, so the force σ_{x2} in the x direction is

$$S_{x2} = -(\sigma_x)_x b \, dy + (\sigma_x)_{x+dx} b \, dy = \left\{ -\sigma_x + \left(\sigma_x + \frac{\partial \sigma_x}{\partial x} dx \right) \right\} b \, dy$$

$$= \frac{\partial \sigma_x}{\partial x} b \, dx \, dy = 2\mu \frac{\partial^2 u}{\partial x^2} b \, dx \, dy \quad (6.10)$$

Therefore,

$$\left. \begin{array}{l} S_x = S_{x1} + S_{x2} = \mu \left(\dfrac{\partial^2 u}{\partial x^2} + \dfrac{\partial^2 u}{\partial y^2} \right) b \, dx \, dy \\[3mm] S_y = \mu \left(\dfrac{\partial^2 v}{\partial x^2} + \dfrac{\partial^2 v}{\partial y^2} \right) b \, dx \, dy \end{array} \right\} \quad (6.11)$$

Louis Marie Henri Navier (1785–1836)
Born in Dijon, France. Actively worked in the educational and bridge engineering fields. His design of a suspension bridge over the River Seine in Paris attracted public attention. In analysing fluid movement, thought of an assumed force by repulsion and absorption between neighbouring molecules in addition to the force studied by Euler to find the equation of motion of fluid. Thereafter, through research by Cauchy, Poisson and Saint-Venant, Stokes derived the present equations, including viscosity.

Substituting eqns (6.7), (6.8) and (6.10) into eqn (6.5), the following equation is obtained:

$$\left.\begin{aligned}
\rho\left(\frac{\partial u}{\partial t}+u\frac{\partial u}{\partial x}+v\frac{\partial u}{\partial y}\right) &= \rho X - \frac{\partial p}{\partial x}+\mu\left(\frac{\partial^2 u}{\partial x^2}+\frac{\partial^2 u}{\partial y^2}\right) \\
\rho\left(\frac{\partial v}{\partial t}+u\frac{\partial v}{\partial x}+v\frac{\partial v}{\partial y}\right) &= \underbrace{\rho Y} - \frac{\partial p}{\partial y}+\mu\left(\frac{\partial^2 v}{\partial x^2}+\frac{\partial^2 v}{\partial y^2}\right)
\end{aligned}\right\} \quad (6.12)$$

$$\underbrace{\hspace{4cm}}_{\text{Inertia term}} \quad \underbrace{\text{Body force}}_{\text{term}} \quad \underbrace{\text{Pressure}}_{\text{term}} \quad \underbrace{\hspace{3cm}}_{\text{Viscous term}}$$

These equations are called the Navier–Stokes equations. In the inertia term, the rates of velocity change with position and

$$\left(u\frac{\partial u}{\partial x}+v\frac{\partial u}{\partial y}\right) \quad \left(u\frac{\partial v}{\partial x}+v\frac{\partial v}{\partial y}\right)$$

and so are called the convective accelerations.

In the case of axial symmetry, when cylindrical coordinates are used, eqns (6.12) become the following equations:

$$\left.\begin{aligned}
\rho\left(\frac{\partial u}{\partial t}+u\frac{\partial u}{\partial x}+v\frac{\partial u}{\partial r}\right) &= \rho X - \frac{\partial p}{\partial x}+\mu\left(\frac{\partial^2 u}{\partial x^2}+\frac{1}{r}\frac{\partial u}{\partial r}+\frac{\partial^2 u}{\partial r^2}\right) \\
\rho\left(\frac{\partial v}{\partial t}+u\frac{\partial v}{\partial x}+v\frac{\partial v}{\partial r}\right) &= \rho R - \frac{\partial p}{\partial r}+\mu\left(\frac{\partial^2 v}{\partial x^2}+\frac{1}{r}\frac{\partial v}{\partial r}-\frac{v}{r^2}+\frac{\partial^2 v}{\partial r^2}\right)
\end{aligned}\right\} \quad (6.13)$$

where R is the r direction component of external force acting on the fluid of unit mass.

The vorticity ζ is

$$\zeta = \frac{\partial v}{\partial x}-\frac{\partial u}{\partial r} \quad (6.14)$$

and the shearing stress is

$$\tau = -\mu\left(\frac{\partial u}{\partial r}+\frac{\partial v}{\partial x}\right) \quad (6.15)$$

The continuity equation (6.3), along with equation (6.15), are convenient for analysing axisymmetric flow in pipes.

Now, omitting the body force terms, eliminating the pressure terms by partial differentiation of the upper equation of eqn (6.12) by y and the lower equation by x, and then rewriting these equations using the equation of vorticity (4.7), the following equation is obtained:

$$\rho\left(\frac{\partial \zeta}{\partial t}+u\frac{\partial \zeta}{\partial x}+v\frac{\partial \zeta}{\partial y}\right)=\mu\left(\frac{\partial^2 \zeta}{\partial x^2}+\frac{\partial^2 \zeta}{\partial y^2}\right) \qquad (6.16)$$

For ideal flow, $\mu = 0$, so the right-hand side of eqn (6.11) becomes zero. Then it is clear that the vorticity does not change in the ideal flow process. This is called the vortex theory of Helmholz.

Now, non-dimensionalise the above using the representative size l and the representative velocity U:

$$x^* = x/l \quad y^* = y/l$$
$$u^* = u/U \quad v^* = v/U$$
$$t^* = tU/l \qquad (6.17)$$
$$\zeta^* = \partial v^*/\partial x^* - \partial u^*/\partial y^*$$
$$Re = \rho Ul/\mu$$

Using these equations rewrite eqn (6.16) to obtain the following equation:

$$\frac{\partial \zeta^*}{\partial t^*}+u^*\frac{\partial \zeta^*}{\partial x^*}+v^*\frac{\partial \zeta^*}{\partial y^*}=\frac{1}{Re}\left(\frac{\partial^2 \zeta^*}{\partial x^{*2}}+\frac{\partial^2 \zeta^*}{\partial y^{*2}}\right) \qquad (6.18)$$

Equation (6.18) is called the vorticity transport equation. This equation shows that the change in vorticity due to fluid motion equals the diffusion of vorticity by viscosity. The term $1/Re$ corresponds to the coefficient of diffusion. Since a smaller Re means a larger coefficient of diffusion, the diffusion of vorticity becomes larger, too.

6.3 Velocity distribution of laminar flow

In the Navier–Stokes equations, the convective acceleration in the inertial term is non-linear[2]. Hence it is difficult to obtain an analytical solution for general flow. The strict solutions obtained to date are only for some special flows. Two such examples are shown below.

6.3.1 Flow between parallel plates

Let us study the flow of a viscous fluid between two parallel plates as shown in Fig. 6.4, where the flow has just passed the inlet length (see Section 7.1)

---

[2] The case where an equation is not a simple equation for the unknown function and its partial differential function is called non-linear.

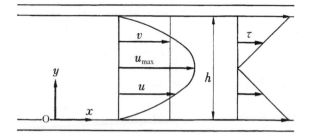

Fig. 6.4 Laminar flow between parallel plates

where it had flowed in the laminar state. For the case of a parallel flow like this, the Navier–Stokes equation (6.12) is extremely simple as follows:

1. As the velocity is only u since $v = 0$, it is sufficient to use only the upper equation.
2. As this flow is steady, u does not change with time, so $\partial u/\partial t = 0$.
3. As there is no body force, $\rho X = 0$.
4. As this flow is uniform, u does not change with position, so $\partial u/\partial x = 0$ and $\partial^2 u/\partial x^2 = 0$.
5. Since $v = 0$, the lower equation of (6.12) simply expresses the hydrostatic pressure variation and has no influence in the x direction.

So, the upper equation of eqn (6.12) becomes

$$\mu \frac{d^2 u}{dy^2} = \frac{dp}{dx} \qquad\qquad (6.19)^3$$

---

[3] Consider the balance of forces acting on the respective faces of an assumed small volume $dx\,dy$ (of unit width) in a fluid.

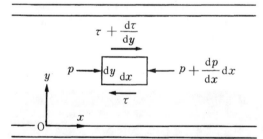

Forces acting on a small volume between parallel plates

Since there is no change of momentum between the two faces, the following equation is obtained:

$$p\,dy - \left(p + \frac{dp}{dx}dx\right)dy - \tau\,dx + \left(\tau + \frac{d\tau}{dy}dy\right)dx = 0$$

Therefore

$$\frac{d\tau}{dy} = \frac{dp}{dx}$$

and

$$\tau = \mu \frac{du}{dy} \quad \text{since} \quad \mu \frac{d^2 u}{dy^2} = \frac{dp}{dx} \qquad\qquad (6.19)'$$

By integrating the above equation twice about y, the following equation is obtained:

$$u = \frac{1}{2\mu}\frac{dp}{dx}y^2 + c_1 y + c_2 \qquad (6.20)$$

Using $u = 0$ as the boundary condition at $y = 0$ and h, c_1 and c_2 are found as follows:

$$u = -\frac{1}{2\mu}\frac{dp}{dx}(h - y)y \qquad (6.21)$$

It is clear that the velocity distribution now forms a parabola.

At $y = h/2$, $du/dy = 0$, so u becomes u_{max}:

$$u_{max} = -\frac{1}{8\mu}\frac{dp}{dx}h^2 \qquad (6.22)$$

The volumetric flow rate Q becomes

$$Q = \int_0^h u\,dy = -\frac{1}{12\mu}\frac{dp}{dx}h^3 \qquad (6.23)$$

From this equation, the mean velocity v is

$$v = \frac{Q}{h} = -\frac{1}{12\mu}\frac{dp}{dx}h^2 = \frac{1}{1.5}u_{max} \qquad (6.24)$$

The shearing stress τ due to viscosity becomes

$$\tau = \mu\frac{du}{dy} = -\frac{1}{2}\frac{dp}{dx}(h - 2y) \qquad (6.25)$$

The velocity and shearing stress distribution are shown in Fig. 6.4.

Figure 6.5 is a visualised result using the hydrogen bubble method. It is clear that the experimental result coincides with the theoretical result.

Putting l as the length of plate in the flow direction and Δp as the pressure difference, and integrating in the x direction, the following relation is obtained:

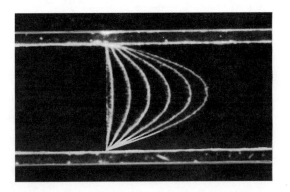

Fig. 6.5 Flow, between parallel plates (hydrogen bubble method), of water, velocity 0.5 m/s, $Re = 140$

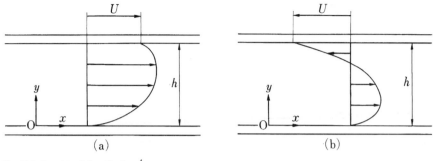

(a)   (b)

Fig. 6.6 Couette–Poiseuille flow[4]

$$-\frac{dp}{dx} = \frac{\Delta p}{l} \tag{6.26}$$

Substituting this equation into eqn (6.23) gives

$$Q = \frac{\Delta p h^3}{12\mu l} \tag{6.27}$$

As shown in Fig. 6.6, in the case where the upper plate moves in the x direction at constant speed U or $-U$, from the boundary conditions of $u = 0$ at $y = 0$ and $u = U$ at $y = h$, c_1 and c_2 in eqn (6.20) can be determined. Thus

$$u = \frac{\Delta p}{2\mu l}(h - y)y \pm \frac{Uy}{h} \tag{6.28}$$

Then, the volumetric flow rate Q is as follows:

$$Q = \int_0^h u \, dy = \frac{\Delta p h^3}{12\mu l} \pm \frac{Uh}{2} \tag{6.29}$$

6.3.2 Flow in circular pipes

A flow in a long circular pipe is a parallel flow of axial symmetry (Fig. 6.7). In this case, it is convenient to use the Navier–Stokes equation (6.13) using cylindrical coordinates. Under the same conditions as in the previous section (6.3.1), simplify the upper equation in equation (6.13) to give

$$\frac{dp}{dx} = \mu \left(\frac{d^2u}{dr^2} + \frac{1}{r}\frac{du}{dr} \right) \tag{6.30}$$

Integrating,

$$u = \frac{1}{4\mu}\frac{dp}{dx}r^2 + c_1 \log r + c_2 \tag{6.31}$$

According to the boundary conditions, since the velocity at $r = 0$ must be finite $c_1 = 0$ and c_2 is determined when $u = 0$ at $r = r_0$:

---

[4] Assume a viscous fluid flowing between two parallel plates; fix one of the plates and move the other plate at velocity U. The flow in this case is called Couette flow. Then, fix both plates, and have the fluid flow by the differential pressure. The flow in this case is called two-dimensional Poiseuille flow. The combination of these two flows as shown in Fig. 6.6 is called Couette–Poiseuille flow.

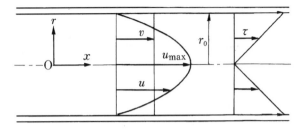

Fig. 6.7 Laminar flow in a circular pipe

$$u = -\frac{1}{4\mu}\frac{dp}{dx}(r_0^2 - r^2) \tag{6.32}$$

From this equation, it is clear that the velocity distribution forms a paraboloid of revolution with u_{max} at $r = 0$:

$$u_{max} = -\frac{1}{4\mu}\frac{dp}{dx}r_0^2 \tag{6.33}$$

The volumetric flow rate passing pipe Q becomes

$$Q = \int_0^{r_0} 2\pi ru\, dr = -\frac{\pi r_0^4}{8\mu}\frac{dp}{dx} \tag{6.34}$$

From this equation, the mean velocity v is

$$v = \frac{Q}{\pi r_0^2} = -\frac{r_0^2}{8\mu}\frac{dp}{dx} = \frac{1}{2}u_{max} \tag{6.35}$$

The shear stress due to the viscosity is,

$$\tau = -\mu\frac{du}{dr} = -\frac{1}{2}\frac{dp}{dx}r \tag{6.36}[5]$$

The velocity distribution and the shear distribution are shown in Fig. 6.7.

---

[5] Equation (6.36) can be deduced by the balance of forces. From the diagram

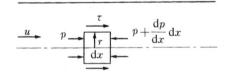

Force acting on a cylindrical element in a round pipe

$$-\pi r^2\frac{dp}{dx} + 2\pi r\tau\, dx = 0$$
$$\tau = \mu\frac{du}{dr}$$

(Since $du/dr < 0$, τ is negative, i.e. leftward.)
Thus

$$\frac{du}{dr} = \frac{1}{2\mu}\frac{dp}{dx}r \tag{6.36$'$}$$

is obtained.

Gotthilf Heinrich Ludwig Hagen (1797–1884)
German hydraulic engineer. Conducted experiments on the relation between head difference and flow rate. Had water mixed with sawdust flow in a brass pipe to observe its flowing state at the outlet. Was yet to discover the general similarity parameter including the viscosity, but reported that the transition from laminar to turbulent flow is connected with tube diameter, flow velocity and water temperature.

A visualisation result using the hydrogen bubble method is shown in Fig. 6.8.

Putting the pressure drop in length l as Δp, the following equation is obtained from eqn (6.33):

$$\Delta p = \frac{128\,\mu l Q}{\pi d^4} = \frac{32\,\mu l v}{d^2} \tag{6.37}$$

This relation was discovered independently by Hagen (1839) and Poiseuille (1841), and is called the Hagen–Poiseuille formula. Using this equation, the viscosity of liquid can be obtained by measuring the pressure drop Δp.

Fig. 6.8 Velocity distribution, in a circular pipe (hydrogen bubble method), of water, velocity 2.4 m/s, $Re = 195$

Jean Louis Poiseuille (1799–1869)
French physician and physicist. Studied the pumping power of the heart, the movement of blood in vessels and capillaries, and the resistance to flow in a capillary. In his experiment on a glass capillary (diameter 0.029–0.142 mm) he obtained the experimental equation that the flow rate is proportional to the product of the difference in pressure by a power of 4 of the pipe inner diameter, and in inverse proportion to the tube length.

6.4 Velocity distribution of turbulent flow

As stated in Section 4.4, flow in a round pipe is stabilised as laminar flow whenever the Reynolds number Re is less than 2320 or so, but the flow becomes turbulent through the transition region as Re increases. In turbulent flow, as observed in the experiment where Reynolds let coloured liquid flow, the fluid particles have a velocity minutely fluctuating in an irregular short cycle in addition to the timewise mean velocity. By measuring the flow with a hot-wire anemometer, the fluctuating velocity as shown in Fig. 6.9 can be recorded.

For two-dimensional flow, the velocity is expressed as follows:

$$u = \bar{u} + u' \qquad v = \bar{v} + v'$$

Fig. 6.9 Turbulence

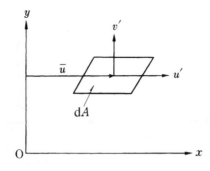

Fig. 6.10 Momentum transport by turbulence

where $\bar{u}$ and $\bar{v}$ are the timewise mean velocities and u' and v' are the fluctuating velocities.

Now, consider the flow at velocity u in the x direction as the flow between two flat plates (Fig. 6.10), so $u = \bar{u} + u'$ but $v = v'$.

The shearing stress τ of a turbulent flow is now the sum of laminar flow shearing stress (viscous friction stress) τ_l, which is the frictional force acting between the two layers at different velocities, and so-called turbulent shearing stress τ_t, where numerous rotating molecular groups (eddies) mix with each other. Thus

$$\tau = \tau_l + \tau_t \tag{6.38}$$

Now, let us examine the turbulent shearing stress only. As shown in Fig. 6.10, the fluid which passes in unit time in the y direction through minute area dA parallel to the x axis is $\rho v' \, dA$. Since this fluid is at relative velocity u', the momentum is $\rho v' \, dA u'$. By the movement of this fluid, the upper fluid increases its momentum per unit area by $\rho u' v'$ in the positive direction of x per unit time. Therefore, a shearing stress develops on face dA. In other words, it is found that the shearing stress due to the turbulent flow is proportional to $\rho u' v'$. Reynolds, by substituting $u = \bar{u} + u'$, $v = \bar{v} + v'$ into the Navier–Stokes equation, performed an averaging operation over time and derived $-\rho \overline{u'v'}$ as a shearing stress in addition to that due to the viscosity. Thus

$$\tau_t = -\rho \overline{u'v'} \tag{6.39}$$

where τ_t is the stress developed by the turbulent flow, which is called the Reynolds stress. As can be seen from this equation, the correlation[6] $\overline{u'v'}$ of

---

[6] In general, the mean of the product of a large enough number of two kinds of quantities is called the correlation. Whenever this value is large, the correlation is said to be strong. In studying turbulent flow, one such correlation is the timewise mean of the products of fluctuating velocities in two directions. Whenever this value is large, it indicates that the velocity fluctuations in two directions fluctuate similarly timewise. Whenever this value is near zero, it indicates that the correlation is small between the fluctuating velocities in two directions. And whenever this value is negative, it indicates that the fluctuating velocities fluctuate in reverse directions to each other.

Ludwig Prandtl (1875–1953)
Born in Germany, Prandtl taught at Hanover Engineering College and then Göttingen University. He successfully observed, by using the floating tracer method, that the surface of bodies is covered with a thin layer having a large velocity gradient, and so advocated the theory of the boundary layer. He is called the creator of modern fluid dynamics. Furthermore, he taught such famous scholars as Blasius and Kármán. Wrote *The Hydrology*.

the fluctuating velocity is necessary for computing the Reynolds stress. Figure 6.11 shows the shearing stress in turbulent flow between parallel flat plates.

Expressing the Reynolds stress as follows as in the case of laminar flow

$$\tau_t = \rho v \frac{d\bar{u}}{dy} \tag{6.40}$$

produces the following as the shearing stress in turbulent flow:

$$\tau = \tau_l + \tau_t = \rho(v + v_t)\frac{d\bar{u}}{dy} \tag{6.41}$$

This v_t is called the turbulent kinematic viscosity. v_t is not the value of a physical property dependent on the temperature or such, but a quantity fluctuating according to the flow condition.

Prandtl assumed the following equation in which, for rotating small parcels

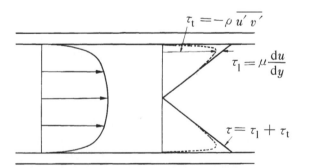

Fig. 6.11 Distribution of shearing stresses of flow between parallel flat plates (enlarged near the wall)

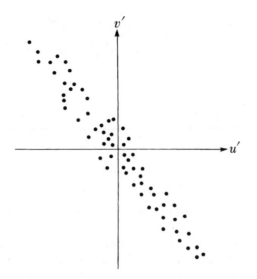

Fig. 6.12 Correlation of u' and v'

of fluid of turbulent flow (eddies) travelling average length, the eddies assimilate the character of other eddies by collisions with them:

$$|u'| \simeq |v'| = l\left|\frac{d\bar{u}}{dy}\right|$$ (6.42)

Prandtl called this l the mixing length.

According to the results of turbulence measurements for shearing flow, the distributions of u' and v' are as shown in Fig. 6.12, where $u'v'$ has a large probability of being negative. Furthermore, the mixing length is redefined as follows, including the constant of proportionality:

$$-\overline{u'v'} = l^2 \left(\frac{d\bar{u}}{dy}\right)^2$$

so that

$$\tau_t = -\rho\overline{u'v'} = \rho l^2 \left(\frac{d\bar{u}}{dy}\right)^2$$ (6.43)[7]

The relation in eqn (6.43) is called Prandtl's hypothesis on mixing length, which is widely used for computing the turbulence shearing stress. Mixing length l is not the value of a physical property but a fluctuating quantity depending on the velocity gradient and the distance from the wall. This

---

[7] According to the convention that the symbol for shearing stress is related to that of velocity gradient, it is described as follows:

$$\tau_t = \rho l^2 \left|\frac{d\bar{u}}{dy}\right|\frac{d\bar{u}}{dy}$$

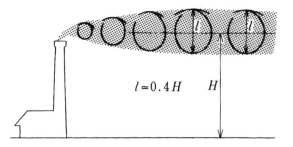

$l \approx 0.4H$ H

Fig. 6.13 Smoke vortices from a chimney

introduction of l is replaced in eqn (6.40) to produce a computable fluctuating quantity.

At this stage, however, Prandtl came to a standstill. That is, unless some concreteness was given to l, no further development could be undertaken. At a loss, Prandtl went outdoors to refresh himself. In the distance there stood some chimneys, the smoke from which was blown by a breeze as shown in Fig. 6.13. He noticed that the vortices of smoke near the ground were not so large as those far from the ground. Subsequently, he found that the size of the vortex was approximately 0.4 times the distance between the ground and the centre of the vortex. On applying this finding to a turbulent flow, he derived the relation $l = 0.4y$. By substituting this relation into eqn (6.43), the following equation was obtained:

$$\frac{d\bar{u}}{dy} = \frac{1}{0.4y}\sqrt{\frac{\tau_t}{\rho}} \qquad (6.44)$$

Next, in an attempt to establish τ_t, he focused his attention on the flow near the wall. There, owing to the presence of wall, a thin layer δ_0 developed where turbulent mixing is suppressed and the effect of viscosity dominates as shown in Fig. 6.14. This extremely thin layer is called the viscous sublayer.[8] Here, the velocity distribution can be regarded as the same as in laminar flow, and v_t in eqn (6.41) becomes almost zero. Assuming τ_0 to be the shearing stress acting on the wall, then so far as this section is concerned:

$$\tau_0 = \mu\frac{du}{dy} = \mu\frac{u}{y} \quad (y \leq \delta_0)$$

or

$$\frac{\tau_0}{\rho} = v\frac{u}{y} \qquad (6.45)$$

$\sqrt{\tau_0/\rho}$ has the dimension of velocity, and is called the friction velocity,

[8] Until some time ago, this layer had been conceived as a laminar flow and called the laminar sublayer, but recently research on visualisation by Kline at Stanford University and others found that the turbulent fluctuation parallel to the wall (bursting process) occurred here, too. Consequently, it is now called the viscous sublayer.

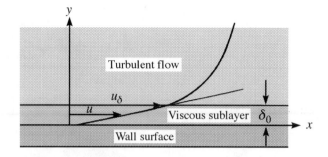

Fig. 6.14 Viscous sublayer

symbol v_* (v star). Substituting, eqn (6.45) becomes:

$$\frac{u}{v_*} = \frac{v_* y}{v}$$ (6.46)

Putting $u = u_\delta$ whenever $y = \delta_0$ gives

$$\frac{u_\delta}{v_*} = \frac{v_* \delta_0}{v} = R_\delta$$ (6.47)

where R_δ is a Reynolds number.

Next, since turbulent flow dominates in the neighbourhood of the wall beyond the viscous sublayer, assume $\tau_t = \tau_0$,[9] and integrate eqn (6.44):

$$\frac{\bar{u}}{v_*} = 2.5 \ln y + c$$ (6.48)

Using the relation $\bar{u} = u_\delta$ when $y = \delta_0$,

$$c = \frac{u_\delta}{v_*} - 2.5 \ln \delta_0 = R_\delta - 2.5 \ln \delta_0$$ (6.49)

Substituting the above into eqn (6.48) gives

$$\frac{\bar{u}}{v_*} = 2.5 \ln \left(\frac{y}{\delta_0}\right) + R_\delta$$

Using the relation in eqn (6.47),

$$\frac{\bar{u}}{v_*} = 2.5 \ln \left(\frac{v_* y}{v}\right) + A$$ (6.50)

If $\bar{u}/v_*$, is plotted against $\log_{10}(v_* y/v)$, it turns out as shown in Fig. 6.15 giving $A = 5.5$.[10]

---

[9] $\tau_t = \tau_0$ was the assumption for the case in the neighbourhood of the wall, and this equation is reasonably applicable when tested off the wall in the direction towards the centre. (Goldstein, S., *Modern Developments in Fluid Dynamics*, (1965), 336, Dover, New York).
[10] It may also be expressed as $\bar{u}/v_* = u^+$, $v_* y/v = y^+$.

Theodor von Kármán (1881–1963)
Studied at the Royal Polytechnic Institute of Budapest, and took up teaching positions at Göttingen University, the Polytechnic Institute of Aachen and California Institute of Technology. Beginning with the study of vortices in the flow behind a cylinder, known as the Kármán vortex street, he left many achievements in fluid dynamics including drag on a body and turbulent flow. Wrote *Aerodynamics: Selected Topics in the Light of Their Historical Development.*

$$\frac{\bar{u}}{v_*} = 5.75 \log\left(\frac{v_* y}{v}\right) + 5.5 \qquad (6.51)$$

This equation is considered applicable only in the neighbourhood of the wall from the viewpoint of its derivation. As seen from Fig. 6.15, however, it was found to be applicable up to the pipe centre from the comparison with the experimental results. This is called the logarithmic velocity distribution, and it is applicable to any value of Re.

In addition, Prandtl separately derived through experiment the following

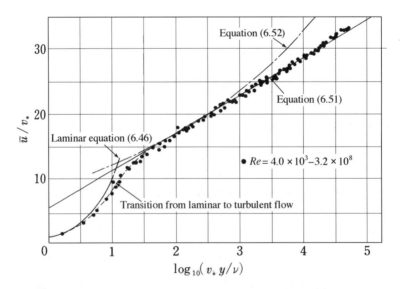

Fig. 6.15 Velocity distribution in a circular pipe (experimental values by Reichardt)

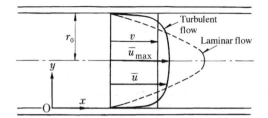

Fig. 6.16 Velocity distribution of turbulent flow

equation of an exponential function as the velocity distribution of a turbulent flow in a circular pipe as shown in Fig. 6.16:

$$\frac{\bar{u}}{\bar{u}_{max}} = \left(\frac{y}{r_0}\right)^{1/n} \quad (0 \leq y \leq r_0) \tag{6.52}$$

n changes according to Re, and is 7 when $Re = 1 \times 10^5$. Since many cases are generally for flows in this neighbourhood, the equation where $n = 7$ is frequently used. This equation is called the Kármán–Prandtl 1/7 power law.[11] Furthermore, there is an experimental equation[12] of $n = 3.45Re^{0.07}$. v/u_{max} is 0.8–0.88

Figure 6.16 also shows the overlaid velocity distributions of laminar and turbulent flows whose average velocities are equal.

Most flows we see daily are turbulent flows, which are important in such applications as heat transfer and mixing. Alongside progress in measuring technology, including visualisation techniques, hot-wire anemometry and laser Doppler velocimetry, and computerised numerical computation, much research is being conducted to clarify the structure of turbulent flow.

6.5 Boundary layer

If the movement of fluid is not affected by its viscosity, it could be treated as the flow of ideal fluid and the viscosity term of eqn (6.11) could be omitted. Therefore, its analysis would be easier. The flow around a solid, however, cannot be treated in such a manner because of viscous friction. Nevertheless, only the very thin region near the wall is affected by this friction. Prandtl identified this phenomenon and had the idea to divide the flow into two regions. They are:

1. the region near the wall where the movement of flow is controlled by the frictional resistance; and
2. the other region outside the above not affected by the friction and, therefore, assumed to be ideal fluid flow.

The former is called the boundary layer and the latter the main flow.

---

[11] Schlichting, H., *Boundary Layer Theory*, (1968), 563, McGraw-Hill, New York.
[12] Itaya, M, *Bulletin of JSME*, 7-26, (1941–2), III-25.

This idea made the computation of frictional drag etc. acting on a body or a channel relatively easy, and thus enormously contributed to the progress of fluid mechanics.

6.5.1 Development of boundary layer

As shown in Fig. 6.17, at a location far from a body placed in a flow, the flow has uniform velocity U without a velocity gradient. On the face of the body the flow velocity is zero with absolutely no slip. For this reason, owing to the effect of friction the flow velocity near the wall varies continuously from zero to uniform velocity. In other words, it is found that the surface of the body is covered by a coat comprising a thin layer where the velocity gradient is large. This layer forms a zone of reduced velocity, causing vortices, called a wake, to be cast off downstream of the body.

We notice the existence of boundary layers daily in various ways. For example, everybody experiences the feeling of the wind blowing (as shown in Fig. 6.18) when standing in a strong wind at the seaside; however, by stretching out on the beach much less wind is felt. In this case the boundary layer on the ground extends to as much as 1 m or more, so the nearer the

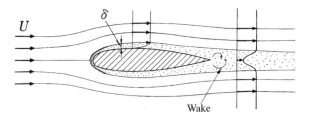

Fig. 6.17 Boundary layer around body

Fig. 6.18 Man lying down is less affected by the coastal breeze than woman standing up

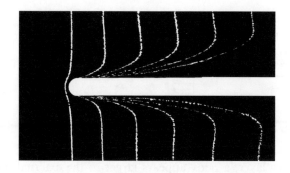

Fig. 6.19 Development of boundary layer on a flat plate (thickness 5 mm) in water, velocity 0.6 m/s

ground the smaller the wind velocity. The velocity u within the boundary layer increases with the distance from the body surface and gradually approaches the velocity of the main flow. Since it is difficult to distinguish the boundary layer thickness, the distance from the body surface when the velocity reaches 99% of the velocity of the main flow is defined as the boundary layer thickness δ. The boundary layer continuously thickens with the distance over which it flows. This process is visualized as shown in Fig. 6.19. This thickness is less than a few millimetres on the frontal part of a high-speed aeroplane, but reaches as much as 50 cm on the rear part of an airship.

When the flow distribution and the drag are considered, it is useful to use the following displacement thickness δ^* and momentum thickness θ instead of δ.

$$U\delta^* = \int_0^\infty (U - u)\mathrm{d}y \qquad (6.53)$$

$$\rho U^2\theta = \rho \int_0^\infty u(U - u)\mathrm{d}y \qquad (6.54)$$

δ^* is the position which equalises two zones of shaded portions in Fig. 6.20(a). It corresponds to an amount δ^* by which, owing to the development

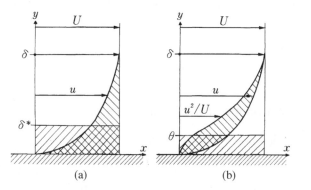

Fig. 6.20 Displacement thickness (a) and momentum thickness (b)

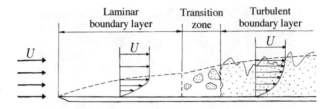

Fig. 6.21 Boundary layer on a flat board surface

of the boundary layer, a body appears larger to the external flow compared with the case where the body is an inviscid fluid. Consequently, in the case where the state of the main flow is approximately obtained as inviscid flow, a computation which assumes the body to be larger by δ^* produces a result nearest to reality. Also, the momentum thickness θ equates the momentum decrease per unit time due to the existence of the body wall to the momentum per unit time which passes at velocity U through a height of thickness θ. The momentum decrease is equivalent to the force acting on the body according to the law of momentum conservation. Therefore the drag on a body generated by the viscosity can be obtained by using the momentum thickness.

Consider the case where a flat plate is placed in a uniform flow. The flow velocity is zero on the plate surface. Since the shearing stress due to viscosity acts between this layer and the layer immediately outside it, the velocity of the outside layer is reduced. Such a reduction extends to a further outside layer and thus the boundary layer increases its thickness in succession, beginning from the front end of the plate as shown in Fig. 6.21.

In this manner, an orderly aligned sheet of vorticity diffuses. Such a layer is called a laminar boundary layer, which, however, changes to a turbulent boundary layer when it reaches some location downstream.

This transition to turbulence is caused by a process in which a very minor disturbance in the flow becomes more and more turbulent until at last it makes the whole flow turbulent. The transition of the boundary layer therefore does not occur instantaneously but necessitates some length in the direction of the flow. This length is called the transition zone. In the transition zone the laminar state and the turbulent state are mixed, but the further the flow travels the more the turbulent state occupies until at last it becomes a turbulent boundary layer.

The velocity distributions in the laminar and turbulent boundary layers are similar to those for the flow in a pipe.

6.5.2 Equation of motion of boundary layer

Consider an incompressible fluid in a laminar boundary layer. Each component of the equation of motion in the y direction is small compared with that in the x direction, while $\partial^2 u/\partial x^2$ is also small compared with $\partial^2 u/\partial y^2$. Therefore, the Navier–Stokes equations (6.12) simplify the following equations:

$$\rho\left(u\frac{\partial u}{\partial x}+v\frac{\partial u}{\partial y}\right)=-\frac{\partial p}{\partial x}+\mu\frac{\partial^2 u}{\partial y^2}\tag{6.55}$$

$$\frac{\partial p}{\partial y}=0\tag{6.56}$$

The continuity equation is as follows:

$$\frac{\partial u}{\partial x}+\frac{\partial v}{\partial y}=0\tag{6.57}$$

Equations (6.55)–(6.57) are called the boundary layer equations of laminar flow.

For a steady-state turbulent boundary layer, with similar considerations, the following equations result:

$$\rho\left(\bar{u}\frac{\partial\bar{u}}{\partial x}+\frac{\partial\bar{u}}{\partial y}\right)=-\frac{\partial\bar{p}}{\partial x}+\frac{\partial\tau}{\partial y}\tag{6.58}$$

$$\tau=\mu\frac{\partial\bar{u}}{\partial y}-\overline{\rho u'v'}\tag{6.59}$$

$$\frac{\partial\bar{p}}{\partial y}=0\tag{6.60}$$

$$\frac{\partial\bar{u}}{\partial x}+\frac{\partial\bar{v}}{\partial y}=0\tag{6.61}$$

Equations (6.58)–(6.61) are called the boundary layer equations of turbulent flow.

6.5.3 Separation of boundary layer

In a flow where the pressure decreases in the direction of the flow, the fluid is accelerated and the boundary layer thins. In a contraction flow, the pressure has such a negative (favourable) gradient that the flow stabilises while the turbulence gradually decreases.

In contrast, things are quite different in a flow with a positive (adverse) pressure gradient where the pressure increases in the flow direction, such as a divergent flow or flow on a curved wall as shown in Fig. 6.22. Fluid far off the wall has a large flow velocity and therefore large inertia too. Therefore, the flow can proceed to a downstream location overcoming the high pressure downstream. Fluid near the wall with a small flow velocity, however, cannot overcome the pressure to reach the downstream location because of its small inertia. Thus the flow velocity becomes smaller and smaller until at last the velocity gradient becomes zero. This point is called the separation point of the flow. Beyond it the velocity gradient becomes negative to generate a flow reversal. In this separation zone, more vortices develop than in the ordinary boundary layer, and the flow becomes more turbulent. For this reason the energy loss increases. Therefore, an expansion flow is readily destabilised with a large loss of energy.

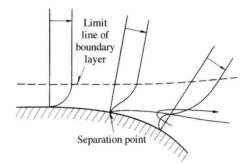

Fig. 6.22 Separation of boundary layer

6.6 Theory of lubrication

As shown in Fig. 6.23, consider two planes with a wedge-like gap containing an oil film between them. Assume that the upper plane is stationary and of length l inclined to the x axis by α, and that the lower plane is an infinitely long plane moving at constant velocity U in the x direction. By the movement of the lower plane the oil stuck to it is pulled into the wedge. As a result, the internal pressure increases to push up the upper plane so that the two planes do not come into contact. This is the principle of a bearing. In this flow, since the oil-film thickness is small in comparison with the length of plane in the flow direction, the flow is laminar where the action of viscosity is very dominant. Therefore, by considering it in the same way as a flow between parallel planes (see Section 6.3.1), the following equation is obtained from eqn (6.12):

$$\frac{\mathrm{d}p}{\mathrm{d}x} = \mu\frac{\partial^2 u}{\partial y^2} \tag{6.62}$$

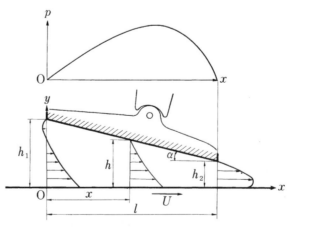

Fig. 6.23 Flow and pressure distribution between inclined planes (slide bearing)

In this case, the pressure p is a function of x only, so the left side is an ordinary differential.

Integrate eqn (6.62) and use boundary conditions $u = U, y = h$ and $u = 0$ at $y = 0$. Then

$$u = U\left(1 - \frac{y}{h}\right) - \frac{dp}{dx}\frac{h^2}{2\mu}\frac{y}{h}\left(1 - \frac{y}{h}\right) \tag{6.63}$$

The flow rate Q per unit width passing here is

$$Q = \int_0^h u\,dy \tag{6.64}$$

Substituting eqn (6.63) into (6.64),

$$Q = \frac{Uh}{2} - \frac{h^3}{12\mu}\frac{dp}{dx} \tag{6.65}$$

From the relation $(h_1 - h_2)/l = \alpha$,

$$h = h_1 - \alpha x \tag{6.66}$$

Substituting the above into eqn (6.65),

$$\frac{dp}{dx} = \frac{6\mu U}{(h_1 - \alpha x)^2} - \frac{12\mu Q}{(h_1 - \alpha x)^3} \tag{6.67}$$

Integrating eqn (6.67),

$$p = \frac{6\mu U}{\alpha(h_1 - \alpha x)} - \frac{6\mu Q}{\alpha(h_1 - \alpha x)^2} + c \tag{6.68}$$

Assume $p = 0$ when $x = 0, x = l$, so

$$Q = \frac{h_1 h_2}{h_1 + h_2} U \quad c = -\frac{6\mu U}{\alpha(h_1 + h_2)}$$

Equation (6.68) becomes as follows:

$$p = \frac{6\mu U(h - h_2)}{(h_1 + h_2)h^2}x \tag{6.69}$$

From eqn (6.69), since $h > h_2$, $p > 0$. Consequently, it is possible to have the upper plane supported above the lower plane. This pressure distribution is illustrated in Fig. 6.23. By integrating this pressure, the supporting load P per unit width of bearing is obtained:

$$P = \int_0^l p\,dx = \frac{6\mu U l^2}{(h_1 - h_2)^2}\left[\log\left(\frac{h_1}{h_2}\right) - 2\frac{h_1 - h_2}{h_1 + h_2}\right] \tag{6.70}$$

From eqn (6.70), the force P due to the pressure reaches a maximum when $h_1/h_2 = 2.2$. At this condition P is as follows:

$$P_{max} = 0.16\frac{\mu U l^2}{h_2^2} \tag{6.71}$$

This slide bearing is mostly used as a thrust bearing. The theory of lubrication above was first analysed by Reynolds.

The principle of the journal bearing is almost the same as the above case. However, since oil-film thickness h is not expressed by the linear equation of x as shown by eqn (6.66), the computation is a little more complicated. This analysis was performed by Sommerfeld and others.

Homer sometimes nods

This is an example in which even such a great figure as Prandtl made a wrong assumption. On one occasion, under the guidance of Prandtl, Hiementz set up a tub to make an experiment for observing a separation point on a cylinder surface. The purpose was to confirm experimentally the separation point computed by the boundary layer theory. Against his expectation, the flow observed in the tub showed violent vibrations.

Hearing of the above vibration, Prandtl responded, 'It was most likely caused by the imperfect circularity of the cylinder section shape.'

Nevertheless, however carefully the cylinder was reshaped, the vibrations never ceased.

Kármán, then an assistant to Prandtl, assumed there was some essential natural phenomenon behind it. He tried to compute the stability of vortex alignment. Summarising the computation over the weekend, he showed the summary to Prandtl on Monday for his criticism. Then, Prandtl told Kármán, 'You did a good job. Make it up into a paper as quickly as possible. I will submit it for you to the Academy.'

A bird stalls

Kármán hit upon the idea of making a bird stall by utilising his knowledge in aerodynamics. When he was standing on the bank of Lake Constance with a piece of bread in his hand, a gull approached him to snatch the bread. Then he slowly withdrew his hand, and the gull tried to slow down its speed for snatching. To do this, it had to increase the lift of its wings by increasing their attack angles. In the course of this, the attack angles probably exceeded their effective limits. Thus the gull sometimes lost its speed and fell (see 'stall', page 164).

Benarl and Kármán

Kármán's train of vortices has been known for so long that it is said to appear on a painting inside an ancient church in Italy. Even before Kármán, however, Professor Henry Benarl (1874–1939) of a French university observed and photographed this train of vortices. Therefore, Benarl insisted on his priority in observing this phenomenon at a meeting on International Applied Dynamics. Kármán responded at the occasion 'I am agreeable to calling Henry Benarl Street in Paris what is called Kármán Street in Berlin and London.' With this joke the two became good friends.

6.7 Problems

1. Show that the continuity equation in the flow of a two-dimensional compressible fluid is as follows:

$$\frac{\partial \rho}{\partial t} + \frac{\partial (\rho u)}{\partial x} + \frac{\partial (\rho v)}{\partial y} = 0$$

2. If the flow of an incompressible fluid is axially symmetric, develop the continuity equation using cylindrical coordinates.

3. If flow is laminar between parallel plates, derive equations expressing (a) the velocity distribution, (b) the mean and maximum velocity, (c) the flow quantity, and (d) pressure loss.

4. If flow is laminar in a circular tube, derive equations expressing (a) the velocity distribution, (b) the mean and maximum velocity, (c) the flow quantity, and (d) pressure loss.

5. If flow is turbulent in a circular tube, assuming a velocity distribution $u = u_{max}(y/r_0)^{1/7}$, obtain (a) the relationship between the mean velocity and the maximum velocity, and (b) the radius of the fluid flowing at mean velocity.

6. Water is flowing at a mean velocity of 4 cm/s in a circular tube of diameter 50 cm. Assume the velocity distribution $u = u_{max}(y/r_0)^{1/7}$. If the shearing stress at a location 5 cm from the wall is $5.3 \times 10^{-3} \mathrm{N/m^2}$, compute the turbulent kinematic viscosity and the mixing length. Assume that the water temperature is 20°C and the mean velocity is 0.8 times the maximum velocity.

7. Consider a viscous fluid flowing in a laminar state through the annular gap between concentric tubes. Derive an equation which expresses the amount of flow in this case. Assume that the inner diameter is d, the gap is h, and $h \ll d$.

8. Oil of 0.09 Pa s (0.9 P) fills a slide bearing with a flat upper face of length 60 cm. A load of 5×10^2 N per 1 cm of width is desired to be supported on the upper surface. What is the maximum oil-film thickness when the lower surface moves at a velocity of 5 m/s?

9. Show that the friction velocity $\sqrt{\tau_0/\rho}$ (τ_0: shearing stress of the wall; ρ: fluid density) has the dimension of velocity.

10. The piston shown in Fig. 6.24 is moving from left to right in a cylinder at a velocity of 6 m/s. Assuming that lubricating oil fills the gap between the piston and the cylinder to produce an oil film, what is the friction force acting on the moving piston? Assume that the kinematic viscosity of oil $v = 50$ cSt, specific gravity $= 0.9$, diameter of cylinder $d_1 = 122$ mm, diameter of piston $d_2 = 125$ mm, piston length $l = 160$ mm, and that the pressure on the left side of the piston is higher than that on the right side by 10 kPa.

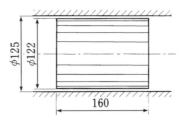

$\phi 125$ $\phi 122$

160

Fig. 6.24

7

Flow in pipes

Consider the flow of an incompressible viscous fluid in a full pipe. In the preceding chapter efforts were made analytically to find the relationship between the velocity, pressure, etc., for this case. In this chapter, however, from a more practical and materialistic standpoint, a method of expressing the loss using an average flow velocity is stated. By extending this approach, studies will be made on how to express losses caused by a change in the cross-sectional area of a pipe, a pipe bend and a valve, in addition to the frictional loss of a pipe.

Lead city water pipe (Roman remains, Bath, England)

Sending water by pipe has a long history. Since the time of the Roman Empire (about 1BC) lead pipes and clay pipes have been used for the water supply system in cities.

7.1 Flow in the inlet region

Consider a case where fluid runs from a tank into a pipe whose entrance section is fully rounded. At the entrance, the velocity distribution is roughly uniform while the pressure head is lower by $v^2/2g$ (v: average flow velocity).

Since the velocity of a viscous fluid is zero on the wall, the fluid near the wall is decelerated. The range subject to deceleration extends as the fluid flows further downstream, until at last the boundary layers develop up to the pipe centre. For this situation, shown in Fig. 7.1, the section from the entrance to just where the boundary layer develops to the tube centre is called the inlet or entrance region, whose length is called the inlet or entrance length. For the value of L, there are the following equations:

Laminar flow:

$$L = 0.065 Red \quad \begin{cases} \text{computation by Boussinesq} \\ \text{experiment by Nikuradse} \end{cases}$$

$$L = 0.06 Red \quad \text{computation by Asao, Iwanami and Mori}$$

Turbulent flow:

$$L = 0.693 Re^{1/4} d \quad \text{computation by Latzko}$$
$$L = (25 \sim 40)d \quad \text{experiment by Nikuradse}$$

Downstream of the inlet region, the static pressure of the pipe line as measured by the liquid column gauge set in the pipe line turns out, as shown in Fig. 7.1, to be lower by H than the water level of the tank, where

$$H = \lambda \frac{l}{d} \frac{v^2}{2g} + \xi \frac{v^2}{2g} \tag{7.1}$$

$\lambda(l/d)(v^2/2g)$ expresses the frictional loss of head (the lost energy of fluid per unit weight). $\xi(v^2/2g)$ expresses the pressure reduction equivalent to the sum of the velocity stored when the velocity distribution is fully developed plus the additional frictional energy loss above that in fully developed flow consumed during the change in velocity distribution.

The velocity energy of the fluid which has attained the fully developed velocity distribution when $x = L$ is

$$E = \int_0^{d/2} 2\pi r u \frac{\rho u^2}{2} dr \tag{7.2}$$

E is calculated by substituting the equations for the velocity distribution for laminar flow (6.32) into u of this equation. The velocity energy for the same flow at the average velocity is

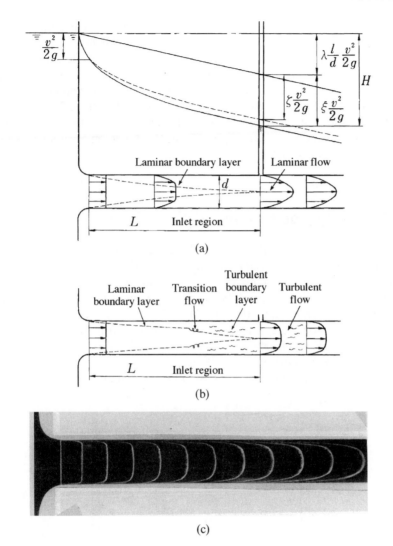

Fig. 7.1 Flow in a circular pipe: (a) laminar flow; (b) turbulent flow; (c) laminar flow (flow visualisation using hydrogen bubble method)

$$E' = \frac{\pi d^2}{4} v \frac{\rho v^2}{2}$$

Putting $E/E' = \zeta$ gives $\zeta = 2$. For the case of turbulent flow, ζ is found to be 1.09 through experiment. ζ is known as the kinetic energy correction factor.

The velocity head equivalent to this energy is

$$\frac{E}{\frac{1}{4}\pi d^2 v \rho g} = \zeta \frac{v^2}{2g} \tag{7.3}$$

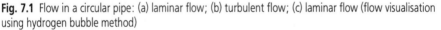

This means that, to compensate for this increase in velocity head when the entrance length reaches L, the pressure head must decrease by the same amount. Furthermore, with the extra energy loss due to the changing velocity distribution included, the value of ξ turns out to be much larger than ζ. $\xi(v^2/2g)$ expresses how much further the pressure would fall than for frictional loss in the inlet region of the pipe if a constant velocity distribution existed. With respect to the value of ξ, for laminar flow values of $\xi = 2.24$ (computation by Boussinesq), 2.16 (computation by Schiller), 2.7 (experiment by Hagen) and 2.36 (experiment by Nakayama and Endo) were reported, while for turbulent flow $\xi = 1.4$ (experiment by Hagen on a trumpet-like tube without an entrance).

7.2 Loss by pipe friction

Let us study the flow in the region where the velocity distribution is fully developed after passing through the inlet region (Fig. 7.2). If a fluid is flowing in the round pipe of diameter d at the average flow velocity v, let the pressures at two points distance l apart be p_1 and p_2 respectively. The relationship between the velocity v and the loss head $h = (p_1 - p_2)/\rho g$ is illustrated in Fig. 7.3, where, for the laminar flow, the loss head h is proportional to the flow velocity v as can clearly be seen from eqn (6.37). For the turbulent flow, it turns out to be proportional to $v^{1.75\sim2}$.

The loss head is expressed by the following equation as shown in eqn (7.1):

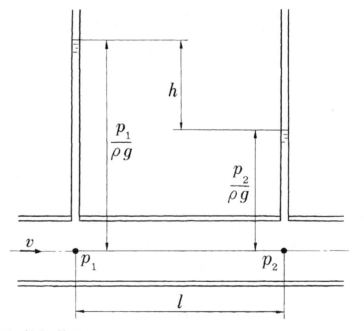

Fig. 7.2 Pipe frictional loss

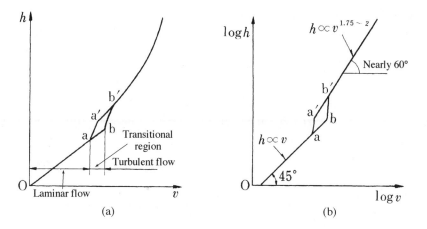

Fig. 7.3 Relationship between flow velocity and loss head

$$h = \lambda \frac{l}{d} \frac{v^2}{2g} \tag{7.4}$$

This equation is called the Darcy–Weisbach equation[1], and the coefficient λ is called the friction coefficient of the pipe.

7.2.1 Laminar flow

In this case, from eqns (6.37) and (7.4),

$$\lambda = 64 \frac{\mu}{\rho v d} = \frac{64}{Re} \tag{7.5}$$

No effect of wall roughness is seen. The reason is probably that the flow turbulence caused by the wall face coarseness is limited to a region near the wall face because the velocity and therefore inertia are small, while viscous effects are large in such a laminar region.

7.2.2 Turbulent flow

λ generally varies according to Reynolds number and the pipe wall roughness.

Smooth circular pipe
The roughness is inside the viscous sublayer if the height ε of wall face ruggedness is

$$\varepsilon \le 5v/v \quad \text{(fluid dynamically smooth)} \tag{7.6}$$

---

[1] In place of λ, many British texts use $4f$ in this equation. Since friction factor $f = \lambda/4$, it is essential to check the definition to which a value of friction factor refers. The symbol used is not a reliable guide.

From eqn (6.45) and Fig. 6.15, no effect of roughness is seen and λ varies according to Reynolds number only; thus the pipe can be regarded as a smooth pipe.

In the case of a smooth pipe, the following equations have been developed:

equation of Blasius: $\lambda = 0.3164 Re^{-1/4}$ $(Re = 3 \times 10^3 \sim 1 \times 10^5)$ (7.7)

equation of Nikuradse:

$$\lambda = 0.0032 + 0.221 Re^{-0.237} \quad (Re = 10^5 \sim 3 \times 10^6) \tag{7.8}$$

equation of Kármán–Nikuradse:

$$\lambda = 1/[2\log_{10}(Re\sqrt{\lambda}) - 0.8]^2 \quad (Re = 3 \times 10^3 \sim 3 \times 10^6) \tag{7.9}$$

equation of Itaya:[2] $\lambda = \dfrac{0.314}{0.7 - 1.65\log_{10}(Re) + (\log_{10} Re)^2}$ (7.10)

By combining eqn (7.4) with (7.7), the relationship $h = cv^{1.75}$ (here c is a constant) arises giving the relationship for turbulent flow in Fig. 7.3.

Rough circular pipe
From eqn (6.51) and Fig. 6.15, where

$$\varepsilon \geq 70v/v_* \quad \text{(fully coarse)} \tag{7.11}$$

the wall face roughness extends into the turbulent flow region. This defines the rough pipe case where λ is determined by the roughness only, and is not related to Reynolds number value.

To simulate regular roughness, Nikuradse performed an experiment in 1933 by lacquer-pasting screened sand grains of uniform diameter onto the inner wall of a tube, and obtained the result shown in Fig. 7.4.

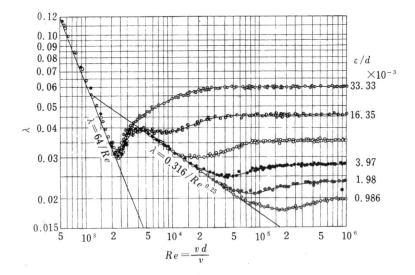

Fig. 7.4 Friction coefficient of coarse circular pipe with sand grains

[2] Itaya, M., *Journal of JSME*, 48 (1945), 84.

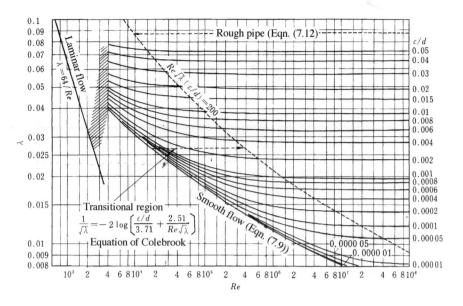

Fig. 7.5 Moody diagram

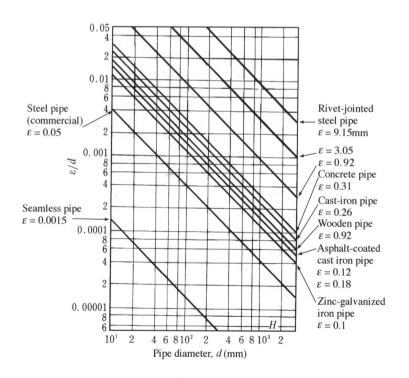

Fig. 7.6 Roughness of commercial pipe

According to this result, whenever $Re > 900(\varepsilon/d)$, it turns out that

$$\lambda = \frac{1}{[1.74 - 2\log_{10}(2\varepsilon/d)]^2} \tag{7.12}$$

The velocity distribution for this case is expressed by the following equation:

$$u/v_* = 8.48 + 5.75\log_{10}(y/\varepsilon) \tag{7.13}$$

For a pipe of irregular coarseness found in practice, the Moody diagram[3] shown in Fig. 7.5 is applicable. For a new commercial pipe, λ can be easily obtained from Fig. 7.5 using ε/d in Fig. 7.6.

7.3 Frictional loss on pipes other than circular pipes

In the case of a pipe other than a circular one (e.g. oblong or oval), how can the pressure loss be found?

Where fluid flows in an oblong pipe as shown in Fig. 7.7, let the pressure drop over length l be h, the sides of the pipe be a and b respectively, and the wall perimeter in contact with the fluid on the section be s, where the shearing stress is τ_0, the shearing force acting on the pipe wall of length l is $l\tau_0 s$, and the balancing pressure force is $\rho g h A$. Then

$$\rho g h A = \tau_0 s l \tag{7.14}$$

This equation shows that for a given pressure loss τ_0 is determined by A/s (the ratio of the flow section area to the wetted perimeter). $A/s = m$ is called the hydraulic mean depth (see Section 8.1). In the case of a filled circular section pipe, since $A = (\pi/4)d^2$, $s = \pi d$, the relationship $m = d/4$ is obtained. So, for pipes other than circular, calculation is made using the following equation and substituting $4m$ (which is called the hydraulic diameter) as the representative size in place of d in eqn (7.4):

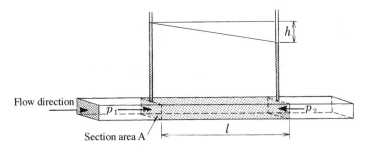

Fig. 7.7 Flow in oblong pipe

---

[3] Moody, L.F. and Princeton, N.J., *Transactions of the ASME*, 66 (1944), 671.

$$h = \lambda \frac{l}{4m} \frac{v^2}{2g} \quad \lambda = f(Re, \varepsilon/4m) \tag{7.15}$$

Here, assuming $Re = 4mv/v$, $\varepsilon/d = \varepsilon/4m$ may be found from the Moody diagram for a circular pipe. Meanwhile, $4m$ is described by the following equations respectively for an oblong section of a by b and for co-axial pipes of inner diameter d_1 and outer diameter d_2:

$$4\frac{ab}{2(a+b)} = \frac{2ab}{a+b} \quad 4\frac{(\pi/4)(d_2^2 - d_1^2)}{\pi(d_1 + d_2)} = d_2 - d_1 \tag{7.16}$$

7.4 Various losses in pipe lines

In a pipe line, in addition to frictional loss, head loss is produced through additional turbulence arising when fluid flows through such components as change of area, change of direction, branching, junction, bend and valve. The loss head for such cases is generally expressed by the following equation:

$$h_s = \zeta \frac{v^2}{2g} \tag{7.17}$$

v in the above equation is the mean flow velocity on a section not affected by the section where the loss head is produced. Where the mean flow velocity changes upstream or downstream of the loss-producing section, the larger of the flow velocities is generally used.

7.4.1 Loss with sudden change of area

Flow expansion
The flow expansion loss h_s for a suddenly widening pipe becomes the following, as already shown by eqn (5.44):

$$h_s = \frac{(v_1 - v_2)^2}{2g} = \left(1 - \frac{A_1}{A_2}\right)^2 \frac{v_1^2}{2g} \tag{7.18}$$

In practice, however, it becomes

$$h_s = \zeta \frac{(v_1 - v_2)^2}{2g} \tag{7.19}$$

or as follows:

$$h_s = \zeta \frac{v_1^2}{2g} \tag{7.20}$$

$$\zeta = \xi \left(1 - \frac{A_1}{A_2}\right)^2 \tag{7.21}$$

Here, ξ is a value near one.

At the outlet of the pipe as shown in Fig. 7.8, since $v_2 = 0$, eqn (7.19) becomes

$$h_s = \xi \frac{v_1^2}{2g} \tag{7.22}$$

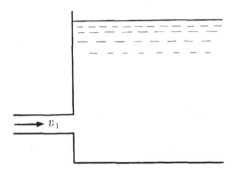

Fig. 7.8 Outlet of pipe line

Flow contraction
Owing to the inertia, section 1 (section area A_1) of the fluid (Fig. 7.9) shrinks to section 2 (section area A_c), and then widens to section 3 (section area A_2). The loss when the flow is accelerated is extremely small, followed by a head loss similar to that in the case of sudden expansion. Like eqn (7.18), it is expressed by

$$h_s = \frac{(v_c - v_2)^2}{2g} = \left(\frac{A_2}{A_c} - 1\right)^2 \frac{v_2^2}{2g} = \left(\frac{1}{C_c} - 1\right)^2 \frac{v_2^2}{2g} \tag{7.23}$$

Here $C_c = A_c/A_2$ is a contraction coefficient. For example, when $A_2/A_1 = 0.1$, $C_c = 0.61$.[4]

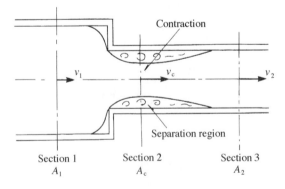

Contraction

v_1

v_c

v_2

Separation region

| Section 1 | Section 2 | Section 3 |
| A_1 | A_c | A_2 |

Fig. 7.9 Sudden contraction pipe

---

[4] Summarised in Donald S. Miller *Internal Flow Systems*, British Hydromechanics Research Association (1978).

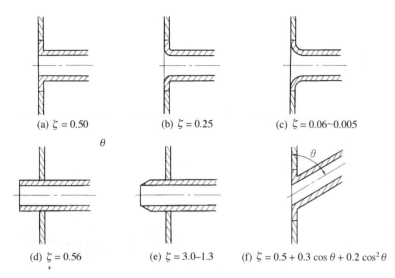

(a) $\zeta = 0.50$ (b) $\zeta = 0.25$ (c) $\zeta = 0.06\!\sim\!0.005$

(d) $\zeta = 0.56$ (e) $\zeta = 3.0\text{–}1.3$ (f) $\zeta = 0.5 + 0.3 \cos \theta + 0.2 \cos^2 \theta$

Fig. 7.10 Inlet shape and loss factor

Inlet of pipe line   As shown in Fig. 7.10, the loss of head in the case where fluid enters from a large vessel is expressed by the following equation:

$$h_s = \zeta \frac{v^2}{2g} \tag{7.24}$$

In this case, however, ζ is the inlet loss factor and v is the mean flow velocity in the pipe. The value of ζ will be the value as shown in Fig. 7.10.[5]

Throttle   A device which decreases the flow area, bringing about the extra resistance in a pipe, is generally called a throttle. There are three kinds of throttle, i.e. choke, orifice and nozzle. If the length of the narrow section is long compared with its diameter, the throttle is called a choke. Since the orifice is explained in Sections 5.2.2 and 11.2.2, and a nozzle is dealt with in Section 11.2.2, only the choke will be explained here.

The coefficient of discharge C in Fig. 7.11 can be expressed as follows, as eqn (5.25), where the difference between the pressure upstream and downstream of the throttle is Δp:

$$Q = C\frac{\pi d^2}{4}\sqrt{\frac{2\Delta p}{\rho}} \tag{7.25}$$

and C is expressed as a function of the choke number $\sigma = Q/vl$. C is as shown in Fig. 7.12, and is expressed by the following equations:[6] if the entrance is

[5] Weisbach, J., *Ingenieur- und Machienen-Mechanik*, I (1896), 1003.
[6] Hibi, *et al.*, *Journal of the Japan Hydraulics & Pneumatics Society*, 2 (1971), 72.

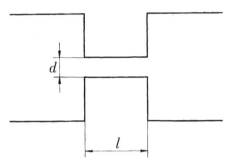

Fig. 7.11 Choke

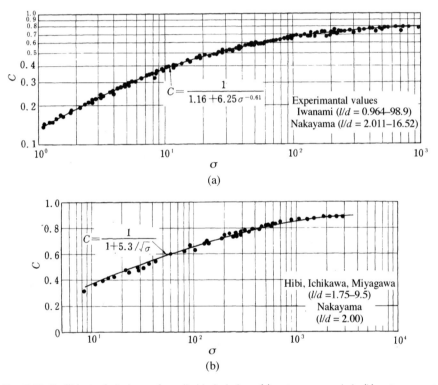

Fig. 7.12 Coefficient of discharge for cylindrical chokes: (a) entrance rounded; (b) entrance not rounded

rounded:

$$C = \frac{1}{1.16 + 6.25\sigma^{-0.61}} \tag{7.26}$$

and if the entrance is not rounded:

$$C = \frac{1}{1 + 5.3/\sqrt{\sigma}} \tag{7.27}$$

7.4.2 Loss with gradual change of area

Divergent pipe or diffuser

The head loss for a divergent pipe as shown in Fig. 7.13 is expressed in the same manner as eqn (7.19) for a suddenly widening pipe:

$$h_s = \xi \frac{(v_1 - v_2)^2}{2g} \tag{7.28}$$

The value of ξ for circular divergent pipes is shown in Fig. 7.14.[7] The value of ξ varies according to θ. For a circular section $\xi = 0.135$ (minimum) when $\theta = 5°30'$. For the rectangular section, $\xi = 0.145$ (minimum) when $\theta = 6°$, and $\xi = 1$ (almost constant) whenever $\theta = 50°-60°$ or more.

For a two-dimensional duct, if θ is small the fluid flows attaches to one of the side walls due to a wall attachment phenomenon (the wall effect).[8] In the case of a circular pipe, when θ becomes larger than the angle which gives the minimum value of ξ, the flow separates midway as shown in Fig. 7.15. Owing to the turbulence accompanying such a separation of flow, the loss of head suddenly increases. This phenomenon is visualised in Fig. 7.16.

A divergent pipe is also used as a diffuser to convert velocity energy into pressure energy. In the case of Fig. 7.13, the following equation is obtained by applying Bernoulli's principle:

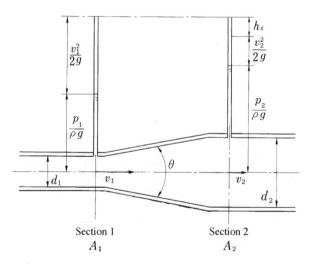

Section 1   Section 2
A_1 A_2

Fig. 7.13 Divergent flow

[7] Gibson, A. H., *Hydraulics*, (1952), 91, Constable, London; Uematsu, T., *Bulletin of JSME*, 2 (1936), 254.
[8] An adjacent wall restricts normal flow entrainment by a jet. A fall in pressure results which deflects the jet such that it can become attached to the wall. This is called the Coanda effect, discovered by H. Coanda in 1932. The effect is the basic principle of the technology of fluidics.

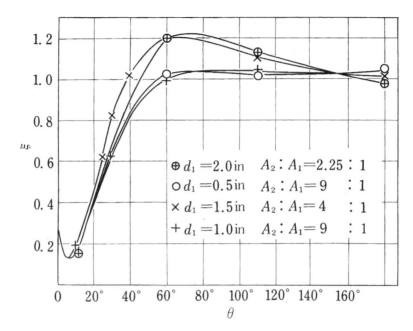

ζₛ

$\oplus\ d_1 = 2.0\,\text{in}\quad A_2 : A_1 = 2.25 : 1$
$\bigcirc\ d_1 = 0.5\,\text{in}\quad A_2 : A_1 = 9\ \ \ : 1$
$\times\ d_1 = 1.5\,\text{in}\quad A_2 : A_1 = 4\ \ \ : 1$
$+\ d_1 = 1.0\,\text{in}\quad A_2 : A_1 = 9\ \ \ : 1$

Fig. 7.14 Loss factor for divergent pipes

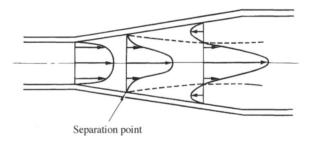

Separation point

Fig. 7.15 Velocity distribution in a divergent pipe

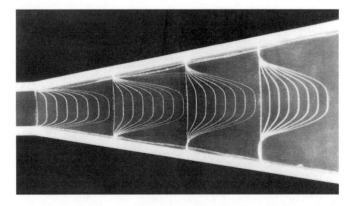

Fig. 7.16 Separation occurring in a divergent pipe (hydrogen bubble method), in water; inlet velocity 6 cm/s, *Re* (inlet port) = 900, divergent angle 20°

$$\frac{p_1}{\rho g} + \frac{v_1^2}{2g} = \frac{p_2}{\rho g} + \frac{v_2^2}{2g} + h_s$$

Therefore

$$\frac{p_2 - p_1}{\rho g} = \frac{v_1^2 - v_2^2}{2g} - h_s \tag{7.29}$$

Putting $p_{2\text{th}}$ for p_2 for the case where there is no loss,

$$\frac{p_{2\text{th}} - p_1}{\rho g} = \frac{v_1^2 - v_2^2}{2g} \tag{7.30}$$

The pressure recovery efficiency η for a diffuser is therefore

$$\eta = \frac{p_2 - p_1}{p_{2\text{th}} - p_1} = 1 - \frac{h_s}{(v_1^2 - v_2^2)/2g} \tag{7.31}$$

Substituting in eqn (7.28), the above equation becomes

$$\eta = 1 - \xi \frac{v_1 - v_2}{v_1 + v_2} = 1 - \xi \frac{1 - A_1/A_2}{1 + A_1/A_2} \tag{7.32}$$

Convergent pipe

In the case where a pipe section gradually becomes smaller, since the pressure decreases in the direction of the flow, the flow runs freely without extra turbulence. Therefore, losses other than the pipe friction are normally negligible.

7.4.3 Loss whenever the flow direction changes

Bend

The gently curving part of a pipe shown in Fig. 7.17 is referred to as a pipe

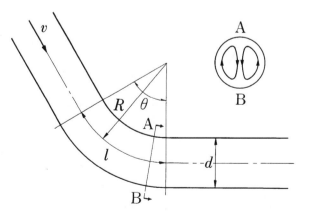

Fig. 7.17 Bend

Table 7.1 Loss factor ζ for bends (smooth wall $Re = 225\,000$, coarse wall face $Re = 146\,000$)

Wall face	$\theta°$	$R/d =$ 1	2	3	4	5
Smooth	15°	0.03	0.03	0.03	0.03	0.03
	22.5°	0.045	0.045	0.045	0.045	0.045
	45°	0.14	0.14	0.08	0.08	0.07
	60°	0.19	0.12	0.095	0.085	0.07
	90°	0.21	0.135	0.10	0.085	0.105
Coarse	90°	0.51	0.51	0.23	0.18	0.20

bend. In a bend, in addition to the head loss due to pipe friction, a loss due to the change in flow direction is also produced. The total head loss h_b is expressed by the following equation:

$$h_b = \zeta_b \frac{v^2}{2g} = \left(\zeta + \lambda \frac{l}{d}\right) \frac{v^2}{2g} \tag{7.33}$$

Here, ζ_b is the total loss factor, and ζ is the loss factor due to the bend effect. The values of ζ are shown in Table 7.1.[9]

In a bend, secondary flow is produced as shown in the figure owing to the introduction of the centrifugal force, and the loss increases. If guide blades are fixed in the bend section, the head loss can be very small.

Elbow

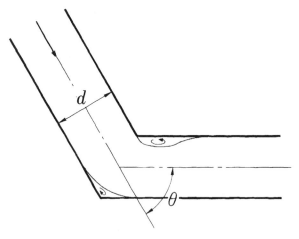

Fig. 7.18 Elbow

[9] Hoffman, A., *Mtt. Hydr. Inst. T. H. München*, 3 (1929), 45; Wasielewski, R. *Mitt, Hydr. Inst. T. H. München*, 5 (1932), 66.

As shown in Fig. 7.18, the section where the pipe curves sharply is called an elbow. The head loss h_b is given in the same form as eqn (7.33). Since the flow separates from the wall in the curving part, the loss is larger than in the case of a bend. Table 7.2 shows values of ζ for elbows.[10]

Table 7.2 Loss factor ζ for elbows

$\theta°$	5°	10°	15°	22.5°	30°	45°	60°	90°
ζ Smooth	0.016	0.034	0.042	0.066	0.130	0.236	0.471	1.129
Coarse	0.024	0.044	0.062	0.154	0.165	0.320	0.687	1.265

7.4.4 Pipe branch and pipe junction

Pipe branch
As shown in Fig. 7.19, a pipe dividing into separate pipes is called a pipe branch. Putting h_{s1} as the head loss produced when the flow runs from pipe ① to pipe ③, and h_{s2} as the head loss produced when the flow runs from pipe ① to pipe ②, these are respectively expressed as follows:

$$h_{s1} = \zeta_1 \frac{v_1^2}{2g} \quad h_{s2} = \zeta_2 \frac{v_1^2}{2g} \tag{7.34}$$

Since the loss factors ζ_1, ζ_2 vary according to the branch angle θ, diameter ratio d_1/d_2 or d_1/d_3 and the discharge ratio Q_1/Q_2 or Q_1/Q_3, experiments were performed for various combinations. Such results were summarised.[11]

Pipe junction
As shown in Fig. 7.20, two pipe branches converging into one are called a pipe junction. Putting h_{s2} as the head loss when the flow runs from pipe ① to pipe ③, and h_{s2} as the head loss when the flow runs from pipe ② to pipe ③, these are expressed as follows:

$$h_{s1} = \zeta_1 \frac{v_3^2}{2g} \quad h_{s2} = \zeta_2 \frac{v_3^2}{2g} \tag{7.35}$$

Values of ζ_1 and ζ_2 are similar to the case of the pipe branch.

[10] Kirchbach, H. und Schubart, W., *Mitt. Hydr. Inst. T. H. München*, 2 (1929), 72; 3 (1929), 121.
[11] Vogel G., *Mitt. Hydr. Inst. T. M. München*, 1 (1926), 75; 2 (1928), 61; Peter-Mann, F., *Mitt. Inst. T. H. München*, 3 (1929), 98.

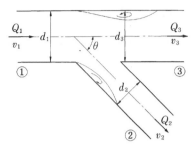

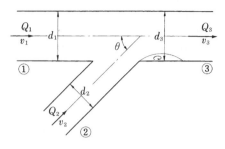

Fig. 7.19 Pipe branch   **Fig. 7.20** Pipe junction

7.4.5 Valve and cock

Head loss on valves is brought about by changes in their section areas, and is expressed by eqn (7.17) provided that v indicates the mean flow velocity at the point not affected by the valve.

Gate valve
The valve as shown in Fig. 7.21 is called a gate valve. Putting d as the diameter and d' as the valve opening, ζ varies according to d'/d. Table 7.3 shows values of ζ for a 1 inch (2.54 cm) nominal diameter valve.[12]

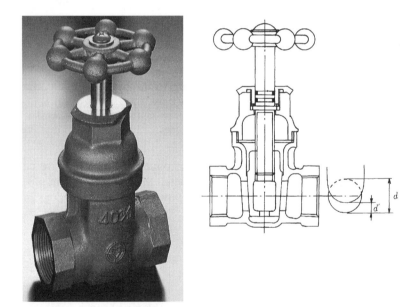

Fig. 7.21 Gate valve

[12] Corp, C.I., *Bulletin of the University of Wisconsin, Engineering Series*, 9-1 (1922), 1.

Globe valve

Table 7.4 shows values of ζ for the globe valve shown in Fig. 7.22, at various openings.[13]

Table 7.3 Values for ζ for 1 inch gate valves ($d = 25.5$ mm)

d'/d	1/8	1/4	3/8	1/2	3/4	1
ζ	211	40.3	10.15	3.54	0.882	0.233

Table 7.4 Values of ζ for 1 inch screw-in globe valves ($d = 25.5$ mm)

l/d	1/4	1/2	3/4	1
ζ	16.3	10.3	7.63	6.09

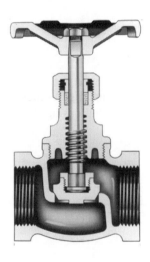
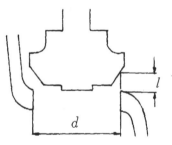

Fig. 7.22 Globe valve

Butterfly valve (Fig. 7.23)

Table 7.5 shows values of ζ for a butterfly valve.[14] As the inclination angle θ of the valve plate increases, the section area immediately downstream of the valve suddenly increases, bringing about an increased value of ζ.

[13] Oki, I., *Suirikigaku* (Hydraulics), 344, Iwanami, Tokyo. In addition, for popet valves, Ichikawa, T. and Shimizu, T., 31 (1965), 317; Kasai, K., *Trans. JSME*, 33 (1967), 1088.
[14] Weisbach, J., *Ingenieur- und Meschienen-Mechanik*, I (1896), 1050.

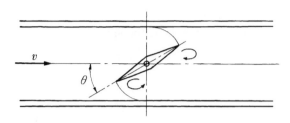

Fig. 7.23 Butterfly valve

Table 7.5 Values of ζ for circular butterfly valves

$\theta°$	10°	20°	30°	50°	70°
z	0.52	1.54	3.91	32.6	751

For a circular butterfly valve, when $\theta = 0°$, the value of ζ is

$$\zeta = t/d \tag{7.36}$$

Cock (Fig. 7.24)

Table 7.6 shows values of ζ for a cock. For cocks, too, as angle θ increases, large changes in section area of flow are brought about, increasing the value of ζ.

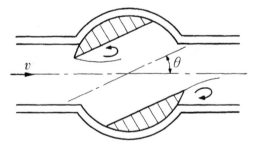

Fig. 7.24 Cock

Table 7.6 Values of ζ for cocks

$\theta°$	10°	30°	50°	60°
ζ	0.29	5.47	52.6	206

Other valves

Values of ζ for various valves are shown in Table 7.7.[15]

Table 7.7 Loss factor for various valves

Valve	Loss coefficients, ζ
Relief valve 	h/d 0.05 0.1 0.15 0.2 0.25 0.3 ζ 3.35 2.85 2.4 2.4 1.7 1.35
Disc valve	Throttle area $a = \pi dx$ Section area of valve seat hole $A = \pi d^2/4$ When $x = d/4$ $a = A$ Loss coefficient $\zeta = 1.3 + 0.2(A/a)^2$
Needle valve	$a = \pi(dx\tan\theta/2 - x^2\tan^2\theta/2)$ $Ax = 0$ when $x = 0$ $\zeta = 0.5 + 0.15(A/a)^2$
Ball valve	$a \simeq 0.75\pi dx$ $\zeta = 0.5 + 0.15(A/a)^2$
Spool valve	At full open position $\zeta = 3 \sim 5.5$

7.4.6 Total loss along a pipe line

For a pipe with flow velocity v, inner diameter d and length l, the total loss from pipe entrance to exit is

$$h = \left(\lambda\frac{l}{d} + \sum\zeta\right)\frac{v^2}{2g}$$ (7.37)

---

[15] Yeaple, F. D., *Hydraulic and Pneumatic Power Control*, (1966), 89, McGraw-Hill, New York.

The first term on the right expresses the total loss by friction, while $\sum \zeta(v^2/2g)$ represents the sum of the loss heads at such sections as the entrance, bend and valve. Whenever a pipe line consists of pipes of different diameters, it is necessary to use the appropriate valve for the flow velocity for each pipe.

When two tanks with a water-level differential h are connected by a pipe line, the exit velocity energy is generally lost. Therefore,

$$h = \left(\lambda \frac{l}{d} + \sum \zeta + 1 \right) \frac{v^2}{2g} \tag{7.38}$$

However, when the pipe line is long such that $l/d > 2000$ and it has no valves of small opening etc., losses other than frictional loss may be neglected.

Conversely, if h is known, the flow velocity could be obtained from eqn (7.37) or eqn (7.38).

In general, for urban water pipes, $v = 1.0 \sim 1.5 \, \text{m/s}$ is typical for long pipe runs, while up to approximately $2.5 \, \text{m/s}$ is typical for short pipe runs. For the headrace of a hydraulic power plant, $2 \sim 5 \, \text{m/s}$ is the usual range.

7.5 Pumping to higher levels

A pump can deliver to higher levels since it gives energy to the water (Fig. 7.25). The head H across the pump is called the total head. The differential

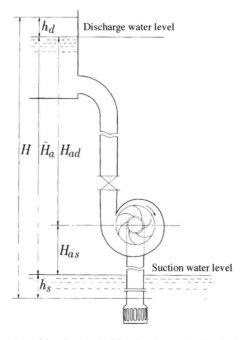

Fig. 7.25 Storage pump: H total head; H_a actual head; $H_{a,s}$ suction head; $H_{s,d}$ discharge head; h_s losses on suction s; h_d losses on discharge side

height H_a between two water levels is called the actual head and

$$H = H_a + h \tag{7.39}$$

where h is the sum of h_s and h_d expressing the total loss.

The volume of water which passes through a pump in unit time is called the pump discharge. Since the energy which a pump gives water in a unit time is H per unit weight, the energy L_w given to water per unit time is

$$L_w = \rho g Q H \tag{7.40}$$

This is sometimes known as the water horsepower.

The power L_s needed by a pump is called the shaft horsepower:

$$L_w / L_s = \eta \tag{7.41}$$

where η is the efficiency of the pump. Since the energy supplied to a pump is not all transmitted to the water due to the energy loss within the pump, it turns out that $\eta < 1$.

As shown in Fig. 7.26, the curve which expresses the relationship between the pump discharge Q and the head H is called the characteristic curve or head curve. In general, the head loss h is proportional to the square of the mean flow velocity in the pipe, and therefore to the square of the pump discharge, and is called the resistance curve. It must be summed with H_a to give the pump load curve.

The pump discharge is given, as shown in Fig. 7.26, by the intersecting point of the head curve and this load curve.

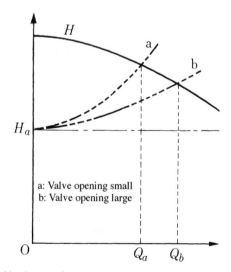

a: Valve opening small
b: Valve opening large

Fig. 7.26 Total head and load curve of pump

7.6 Problems

1. Verify that the kinetic energy for laminar flow in a circular pipe with a fully developed velocity distribution is twice that with uniform velocity.

2. What is the relationship between the flow velocity and the pressure loss in a circular pipe?

3. For laminar flow in a circular pipe, verify that the pipe frictional coefficient can be expressed by the following equation:

$$\lambda = 64/Re$$

4. For turbulent flow in a circular pipe, show that, assuming the pipe frictional coefficient is subject to $\lambda = 0.3164 Re^{-1/4}$, the pressure loss is proportional to a power of 1.75 of the mean flow velocity.

5. For flow in a circular pipe, with constant pipe friction coefficient, show that the frictional head loss is inversely proportional to the fifth power of the pipe diameter. Also, if the diameter is measured with $\alpha\%$ error, what would be the percentage error in head loss?

6. How much head loss will be produced by sending $0.5\,\text{m}^3/\text{min}$ of water a distance of 2000 m using commercial steel pipes of diameter 50 mm? Also, what would be the head loss if the diameter is 100 mm? The water temperature is assumed to be 20°C.

7. What is the necessary shaft horsepower to send $1\,\text{m}^3/\text{min}$ of water through a conduit 100 mm in diameter as shown in Fig. 7.27? Assume pump efficiency $\eta = 80\%$, loss coefficient of sluice valve $\zeta_v = 0.175$, of 90° elbow $\zeta_{90} = 1.265$, of 45° elbow $\zeta_{45} = 0.320$, and pipe frictional coefficient $\lambda = 0.026$.

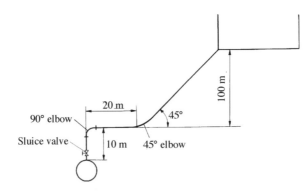

Fig. 7.27

8. A flow of $0.6\,\text{m}^3/\text{s}$ of air discharges through a square duct of sides 20 cm. What is the pressure loss if the duct length is 50 m? Assume an air temperature of 20°C, standard atmospheric pressure, and smooth walls for the duct.

9. Water flows through a sudden expansion where a circular pipe of 40 mm diameter is directly connected to one of 80 mm. If the discharge is $0.08 \, \text{mm}^3/\text{min}$, find the expansion loss.

10. Obtain the head loss and the pressure recovery rate when a circular pipe of 40 mm diameter is connected to one of 80 mm diameter by a $10°$ diffuser.

8

Flow in a water channel

A flowing stream of water where the flow has a free surface exposed to the open air is called a water channel. Included in the water channels, for example, are sewers. Roman waterworks were completed in 302BC with a water channel as long as 16.5 km. In AD305, 14 aqueducts were built with their water channels extending to 578 km in total, it is said. Anyway, water channels have a long history. Figures 1.1 and 8.1 show some remains.

Water channels have such large hydraulic mean depths that the Reynolds numbers are large too. Consequently the flow is turbulent. Furthermore, at such large Reynolds number, the friction coefficient becomes constant and is determined by the roughness of the wall.

Fig. 8.1 Remains of Roman aqueduct

8.1 Flow in an open channel with constant section and flow velocity

In an open channel, the flowing water has a free surface and flows by the action of gravity. As shown in Fig. 8.2, assume that water flows with constant

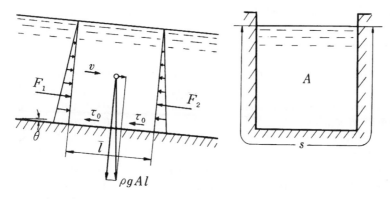

Fig. 8.2 Open channel

velocity v in an open channel of constant section and inclination angle θ of the bottom face. Now examine the balance of forces on water between the two sections a distance l apart. Since the water depth is uniform, the forces F_1 and F_2 acting on the sections due to hydrostatic pressure balance each other. Therefore, the only force acting in the direction of the flow is that component of water weight. Since the flow is not accelerating this force must equal the frictional force due to the wall. If the cross-sectional area of the open channel is A, the length of wetted perimeter s, and the mean value of wall shearing stress τ_0, then

$$\rho g A l \sin \theta = \tau_0 s l$$

Since θ is very small,

$$\text{inclination } i = \tan \theta \simeq \sin \theta$$

Then

$$\tau_0 = \rho g \frac{A}{s} i = \rho g m i \tag{8.1}$$

Here, $m = A/s$ is the hydraulic mean depth.

Expressing τ_0 as $\tau_0 = f \rho v^2/2$ using the frictional coefficient[1] f, then

$$v = \sqrt{\frac{2g}{f} m i} \tag{8.2}$$

Chezy expressed the velocity by the following equation as it was proportional to $\sqrt{mi}$:

$$v = c\sqrt{mi} \tag{8.3}$$

This equation is called Chezy's formula, with c the flow velocity coefficient. The value of c can be obtained using the Ganguillet–Kutter equation:

$$c = \frac{23 + 1/n + 0.00155/i}{1 + [23 + (0.00155/i)](n/\sqrt{m})} \tag{8.4}$$

---

[1] Note that $f = \lambda/4$ (see eqn (7.4)).

Table 8.1 Values of n in the Ganguillet–Kutter, the Manning and α in the Bazin equations

Wall surface condition	n	α
Smoothly shaved wooden board, smooth cement coated	0.010–0.013	0.06
Rough wooden board, relatively smooth concrete	0.012–0.018	
Brick, coated with mortar or like, ashlar masonry	0.013–0.017	0.46
Non-finished concrete	0.015–0.018	
Concrete with exposed gravel	0.016–0.020	1.30
Rough masonry	0.017–0.030	
Both sides stone-paved but bottom face irregular earth	0.028–0.035	
Deep, sand-bed river whose cross-sections are uniform	0.025–0.033	
Gravel-bed river whose cross-sections are uniform and whose banks are covered with wild grass	0.030–0.040	
Bending river with large stones and wild grass	0.035–0.050	2.0

It is also obtainable from the Bazin equation:

$$c = \frac{87}{1 + \alpha/\sqrt{m}} \tag{8.5}$$

More recently the Manning equation has often been used:

$$v = \frac{1}{n} m^{2/3} i^{1/2} \tag{8.6}$$

n in eqns (8.4) and (8.6) as well as α in eqn (8.5) are coefficients varying according to the wall condition. Their values are shown in Table 8.1. In general, the flow velocity is 0.5–3 m/s. These equations and the values appearing in Table 8.1 are for the case of SI units (units m, s).

The discharge of a water channel can be computed by the following equation:

$$Q = Av = Ac\sqrt{mi} = \frac{1}{n} Am^{2/3} i^{1/2} \tag{8.7}$$

The flow velocities at various points of the cross-section are not uniform. The largest flow velocity is found to be $10 \sim 40\%$ of the depth below the water surface, while the mean flow velocity v is at $50 \sim 70\%$ depth.

8.2 Best section shape of an open channel

If the section area A of the flow in an open channel is constant, and given that c and i in eqn (8.3) are also constant, if the section shape is properly selected so that the wetted perimeter is minimised, both the mean flow velocity v and the discharge Q become maximum.

Of all geometrical shapes, if fully charged, a circle has the shortest length of wetted perimeter for the given area. Consequently, a round water channel is important.

8.2.1 Circular water channel

Consider the relationship between water level, flow velocity and discharge for a round water channel of inner radius r (Fig. 8.3).
From eqns (8.6) and (8.7)

$$v = \frac{1}{n}\left(\frac{A}{s}\right)^{2/3} i^{1/2} \qquad Q = \frac{1}{n}\frac{A^{5/3}}{s^{2/3}} i^{1/2}$$

$$A = r^2\left(\frac{\theta}{2}\right) - r^2 \cos\left(\frac{\theta}{2}\right)\sin\left(\frac{\theta}{2}\right) = \frac{r^2(\theta - \sin\theta)}{2}$$

$$s = r\theta$$

$$m = \frac{r}{2}\left(1 - \frac{\sin\theta}{\theta}\right)$$

i.e.

$$v = \frac{1}{n} i^{1/2}\left[\frac{r}{2}\left(1 - \frac{\sin\theta}{\theta}\right)\right]^{2/3} \qquad (8.8)$$

$$Q = \frac{1}{n} i^{1/2}\frac{\theta r^{8/3}}{2^{5/3}}\left(1 - \frac{\sin\theta}{\theta}\right)^{5/3} \qquad (8.9)$$

Putting v_{full} and Q_{full} respectively as the flow velocity and the discharge whenever the maximum capacity of channel is flowing,

$$\frac{v}{v_{full}} = \left(1 - \frac{\sin\theta}{\theta}\right)^{2/3} \qquad (8.10)$$

$$\frac{Q}{Q_{full}} = \frac{\theta}{2\pi}\left(1 - \frac{\sin\theta}{\theta}\right)^{5/3} \qquad (8.11)$$

The relationship between θ and v, Q, is shown in Fig. 8.4.

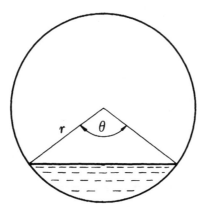

Fig. 8.3 Circular water channel

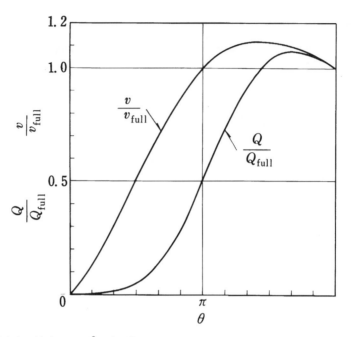

Fig. 8.4 Relationship between θ and v, Q

8.2.2 Rectangular water channel

For the case of Fig. 8.5, obtain the section shape where s is a minimum:

$$s = B + 2H = \frac{A}{H} + 2H$$

$$\frac{\mathrm{d}s}{\mathrm{d}H} = -\frac{A}{H^2} + 2 = 0$$

$$A = 2H^2$$

Therefore,

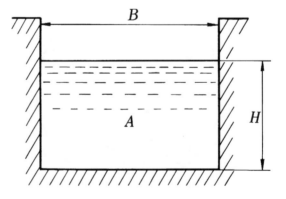

Fig. 8.5 Rectangular water channel

$$\frac{H}{B} = \frac{1}{2}$$

In other words, when c, A and i are constant, in order to maximise v and Q, the depth of the water channel should be one-half of the width.

8.3 Specific energy

Many open channel problems can be solved using the equation of energy. If the pressure is p at a point A in the open channel in Fig. 8.6, the total head of fluid at Point A is

$$\text{total head} = \frac{v^2}{2g} + \frac{p}{\rho g} + z + z_0$$

If the depth of water channel is h, then

$$h = \frac{p}{\rho g} + z$$

Consequently, the total head may be described as follows:

$$\text{total head} = \frac{v^2}{2g} + h + z_0 \qquad (8.12)$$

However, the total head relative to the channel bottom is called the specific energy E, which expresses the energy per unit weight, and if the cross-sectional area of the open channel is A and the discharge Q, then

$$E = h + \frac{Q^2}{2gA^2} \qquad (8.13)$$

This relationship is very important for analysing the flow in an open channel.

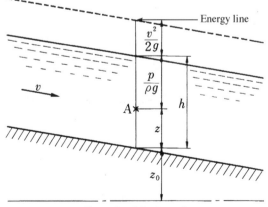

Fig. 8.6 Open channel

There are three variables, E, h, Q. Keeping one of them constant gives the relation between the other two.

8.4 Constant discharge

For constant discharge Q, the relation between the specific energy and the water depth is as shown in Fig. 8.7. The critical point of minimum energy occurs where $dE/dh = 0$.

$$\frac{dE}{dh} = 1 - \frac{Q^2}{gA^3}\frac{dA}{dh} = 0$$

Then

$$\frac{dA}{dh} = \frac{gA^3}{Q^2}$$

When the channel width at the free surface is B, $dA = B\,dh$. So the critical area A_c and the critical velocity v_c become as follows.

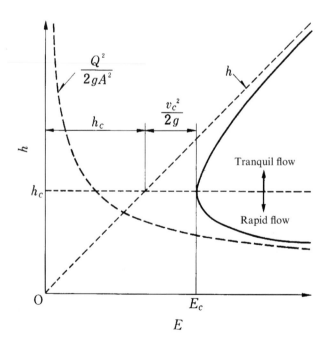

Fig. 8.7 Curve for constant discharge

$$A_c = \left(\frac{BQ^2}{g}\right)^{1/3}$$

$$v_c = \frac{Q}{A_c} = \left(\frac{gA_c}{B}\right)^{1/2}$$

(8.14)

Taking the rectangular water channel as an example, when the discharge per unit width is q, $Q = qB$. As the sectional area $A = hB$, the water depth h_c, eqn (8.15), which makes the specific energy minimum, is obtained from eqn (8.14).

$$h_c = \left(\frac{q^2}{g}\right)^{1/3}$$

(8.15)

At the critical water depth h_c,

$$E_c = \frac{q^2}{2gh_c^2} + h_c$$

From eqn (8.15)

$$q^2 = gh_c^3$$

$$E_c = \frac{h_c}{2} + h_c = 1.5h_c$$

(8.16)

The specific energy (total head) in the critical situation E_c is thus 1.5 times the critical water depth h_c. The corresponding critical velocity v_c becomes, as follows from eqn (8.15),

$$v_c = \frac{q}{h_c} = \sqrt{gh_c}$$

(8.17)

In the critical condition, the flow velocity coincides with the travelling velocity of a wave in a water channel of small depth, a so-called long wave. If the flow depth is deeper or shallower than h_c, the flow behaviour is different. When the water is deeper than h_c, the velocity is smaller than the travelling velocity of the long wave and the flow is called tranquil (or subcritical) flow. When the water is shallower than h_c, the velocity is larger than the travelling velocity of the long wave and the flow is called rapid (or supercritical) flow.

8.5 Constant specific energy

For the case of the rectangular water channel, from eqn (8.13).

$$q^2 = 2g(h^2 E - h^3)$$

$$\frac{dq}{dh} = \frac{g}{q}(2Eh - 3h^2) = 0$$

$$E_c = 1.5h_c$$

(8.18)

Comparing eqn (8.16) with (8.18), both the situation where the discharge is constant while the specific energy is minimum and that where the specific energy is constant while the discharge is maximum are found to be the same (Fig. 8.8).

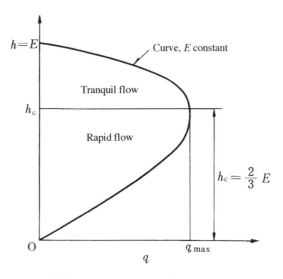

Fig. 8.8 Curve for constant specific energy

8.6 Constant water depth

For the case of the rectangular water channel, from eqn (8.13).

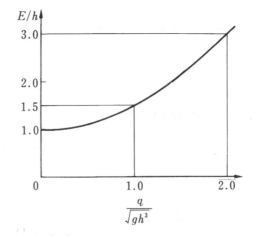

Fig. 8.9 Curve for constant water depth

$$\frac{E}{h} = 1 + \frac{q^2}{2gh^3} \tag{8.19}$$

The relationship between $q/\sqrt{gh^3}$ and E/h is plotted in Fig. 8.9. In words, the specific energy increases parabolically from 1 with q and, when the water depth is critical, i.e. $q^2 = gh^3$, $E/h = 1.5$.

8.7 Hydraulic jump

Rapid flow is unstable, and if decelerated it suddenly shifts to tranquil flow. This phenomenon is called hydraulic jump. For example, as shown in Fig. 8.10(a), when the inclination of a dam bottom is steep, the flow is rapid. When the inclination becomes gentle downstream, the flow is unable to maintain rapid flow and suddenly shifts to tranquil flow. A photograph of this situation is shown in Fig. 8.11.

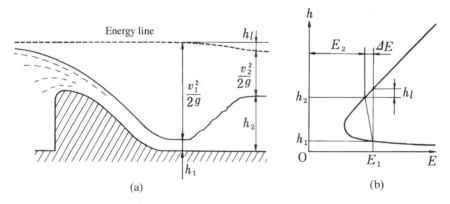

Fig. 8.10 Hydraulic jump

Fig. 8.11 Rapid flow and hydraulic jump on a dam

The travelling velocity a of a long wave in a water channel of small depth h is $\sqrt{gh}$. The ratio of the flow velocity to the wave velocity is called the Froude number. The Froude number of a tranquil flow is less than one, i.e. the flow velocity is smaller than the wave velocity. On the other hand, the Froude number of a rapid flow is larger than one; in other words, the flow velocity is larger than the wave velocity. Thus, tranquil flow and rapid flow in a water channel correspond to subsonic and supersonic flow, respectively, of a compressible gas.

For the flow of gas in a convergent–divergent nozzle (see Section 13.5.3), supersonic flow which has gone through the nozzle stays supersonic if the back pressure is low. If the back pressure is high, however, the flow suddenly shifts to the subsonic flow with a shock wave. In other words, there is an analogy between the hydraulic jump and the shock wave.

When a hydraulic jump is brought about, energy is dissipated by it (Fig. 8.10(b)). Thus erosion of the channel bottom further downstream can be prevented.

8.8 Problems

1. It is desired to obtain $0.5\,\mathrm{m^3/s}$ water discharge using a wooden open channel with a rectangular section as shown in Fig. 8.12. Find the necessary inclination using the Manning equation with $n = 0.01$.

2. For a concrete-coated water channel with the cross-section shown in Fig. 8.13, compare the discharge when the channel inclination is 0.002 obtained by the Chezy and the Manning equation. Assume $n = 0.016$.

3. Find the discharge in a smooth cement-coated rectangular channel 5 m wide, water depth 2 m and inclination 1/2000 using the Bazin equation.

4. Water is sent along the circular conduit in Fig. 8.14. What is the angle θ and depth h which maximise the flow velocity and the discharge if the radius $r = 1.5\,\mathrm{m}$?

5. In an open channel with a rectangular section 3 m wide, $15\,\mathrm{m^3/s}$ of water is flowing at 1.2 m depth. Is the flow rapid or tranquil and what is the specific energy?

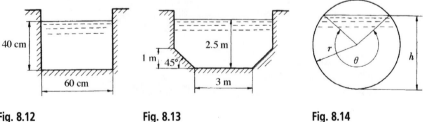

Fig. 8.12   Fig. 8.13   Fig. 8.14

William Froude (1810–79)
Born in England and engaged in shipbuilding. In his sixties started the study of ship resistance, building a boat testing pool (approximately 75 m long) near his home. After his death, this study was continued by his son, Robert Edmund Froude (1846–1924). For similarity under conditions of inertial and gravitational forces, the non-dimensional number used carries his name.

6. Find the critical water depth and the critical velocity when $12 \, \text{m}^3/\text{s}$ of water is flowing in an open channel with a rectangular section 4 m wide.

7. What is the maximum discharge for 2 m specific energy in an open channel with a rectangular section 5 m wide?

8. Water is flowing at $20 \, \text{m}^3/\text{s}$ in a rectangular channel 5 m wide. Find the downstream water depth necessary to cause this flow to jump to tranquil flow.

9. In what circumstances do the phenomena of rapid flow and hydraulic jump occur?

9

Drag and lift

In Chapters 7 and 8 our study concerned 'internal flow' enclosed by solid walls. Now, how shall we consider such cases as the flight of a baseball or golf ball, the movement of an automobile or when an aircraft flies in the air, or where a submarine moves under the water? Here, flows outside such solid walls, i.e. 'external flows', are discussed.

9.1 Flows around a body

Generally speaking, flow around a body placed in a uniform flow develops a thin layer along the body surface with largely changing velocity, i.e. the boundary layer, due to the viscosity of the fluid. Furthermore, the flow separates behind the body, discharging a wake with eddies. Figure 9.1 shows the flows around a cylinder and a flat plate. The flow from an upstream point a is stopped at point b on the body surface with its velocity decreasing to zero; b is called a stagnation point. The flow divides into the upper and lower flows at point b. For a cylinder, the flow separates at point c producing a wake with eddies.

Let the pressure upstream at a, which is not affected by the body, be p_∞, the flow velocity be U and the pressure at the stagnation point be p_0. Then

$$p_0 = p_\infty + \frac{\rho U^2}{2} \tag{9.1}$$

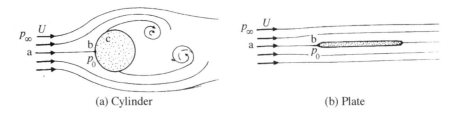

(a) Cylinder   (b) Plate

Fig. 9.1 Flow around a body

9.2 Forces acting on a body

Whenever a body is placed in a flow, the body is subject to a force from the surrounding fluid. When a flat plate is placed in the flow direction, it is only subject to a force in the downstream direction. A wing, however, is subject to the force R inclined to the flow as shown in Fig. 9.2. In general, the force R acting on a body is resolved into a component D in the flow direction U and the component L in a direction normal to U. The former is called drag and the latter lift.

Drag and lift develop in the following manner. In Fig. 9.3, let the pressure of fluid acting on a given minute area dA on the body surface be p, and the friction force per unit area be τ. The force $p\,dA$ due to the pressure p acts normal to dA, while the force due to the friction stress τ acts tangentially. The drag D_p, which is the integration over the whole body surface of the component in the direction of the flow velocity U of this force $p\,dA$, is called form drag or pressure drag. The drag D_f is the similar integration of $\tau\,dA$ and is called the friction drag. D_p and D_f are shown as follows in the form of equations:

$$D_p = \int_A p\,dA \cos\theta \tag{9.2}$$

$$D_f = \int_A \tau\,dA \sin\theta \tag{9.3}$$

The drag D on a body is the sum of the pressure drag D_p and friction drag D_f, whose proportions vary with the shape of the body. Table 9.1 shows the contributions of D_p and D_f for various shapes. By integrating the component of $p\,dA$ and $\tau\,dA$ normal to U, the lift L is obtained.

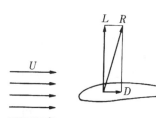

Fig. 9.2 Drag and lift

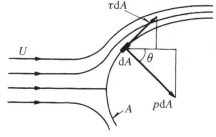

Fig. 9.3 Force acting on body

9.3 The drag of a body

9.3.1 Drag coefficient

The drag D of a body placed in the uniform flow U can be obtained from eqns (9.2) and (9.3). This theoretical computation, however, is generally difficult except for bodies of simple shape and for a limited range of velocity.

Table 9.1 Contributions of D_p and D_f for various shapes

Shape	Pressure drag D_p (%)	Friction drag D_f (%)
	0	100
	≈ 10	≈ 90
	≈ 90	≈ 10
	100	0

Therefore, there is no other way but to rely on experiments. In general, drag D is expressed as follows:

$$D = C_D A \frac{\rho U^2}{2} \qquad (9.4)$$

where A is the projected area of the body on the plane vertical to the direction of the uniform flow and C_D is a non-dimensional number called the drag coefficient. Values of C_D for bodies of various shape are given in Table 9.2.

9.3.2 Drag for a cylinder

Ideal fluid
Let us theoretically study (neglecting the viscosity of fluid) a cylinder placed in a flow. The flow around a cylinder placed at right angles to the flow U of an ideal fluid is as shown in Fig. 9.4. The velocity v_θ at a given point on the cylinder surface is as follows (see Section 12.5.2):

$$v_\theta = 2U \sin \theta \qquad (9.5)$$

Putting the pressure of the parallel flow as p_∞, and the pressure at a given point on the cylinder surface as p, Bernoulli's equation produces the following result:

$$p_\infty + \frac{\rho U^2}{2} = p + \frac{\rho v_\theta^2}{2}$$

$$p - p_\infty = \frac{\rho(U^2 - v_\theta^2)}{2} = \frac{\rho U^2}{2}(1 - 4\sin^2 \theta)$$

$$\frac{p - p_\infty}{\rho U^2 / 2} = 1 - 4\sin^2 \theta \qquad (9.6)$$

Table 9.2 Drag coefficients for various bodies

Body	Dimensional ratio	Datum area, A	Drag coefficient, C_D
Cylinder (flow direction)	$l/d = 1$		0.91
	2		0.85
	4	$\frac{\pi}{4}d^2$	0.87
	7		0.99
Cylinder (right angles to flow)	$l/d = 1$		0.63
	2		0.68
	5		0.74
	10	dl	0.82
	40		0.98
	∞		1.20
Oblong board (right angles to flow)	$a/b = 1$		1.12
	2		1.15
	4		1.19
	10	ab	1.29
	18		1.40
	∞		2.01
Hemisphere (bottomless)	I	$\frac{\pi}{4}d^2$	0.34
	II		1.33
Cone	$a = 60°$	$\frac{\pi}{4}d^2$	0.51
	$a = 30°$		0.34
		$\frac{\pi}{4}d^2$	1.2
Ordinary passenger car	Front projection area		0.28–0.37

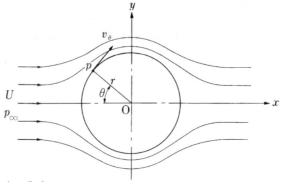

Fig. 9.4 Flow around a cylinder

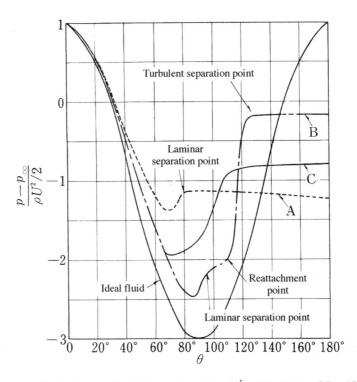

Fig. 9.5 Pressure distribution around cylinder: A, $Re = 1.1 \times 10^5 < Re_c$; B, $Re = 6.7 \times 10^5 > Re_c$; C, $Re = 8.4 \times 10^6 > Re_c$

This pressure distribution is illustrated in Fig. 9.5, where there is left and right symmetry to the centre line at right angles to the flow. Consequently the pressure resistance obtained by integrating this pressure distribution turns out to be zero, i.e. no force at all acts on the cylinder. Since this phenomenon is contrary to actual flow, it is called d'Alembert's paradox, after the French physicist (1717–83).

Viscous fluid

For a viscous flow, behind the cylinder, for very low values of $Re < 1$ ($Re = Ud/v$), the streamlines come together symmetrically as at the front of the cylinder, as indicated in Fig. 9.4. If Re is increased to the range $2 \sim 30$ the boundary layer separates symmetrically at position a (Fig. 9.6(a)) and two eddies are formed rotating in opposite directions.[1] Behind the eddies, the main streamlines come together. With an increase of Re, the eddies elongate and at $Re = 40 \sim 70$ a periodic oscillation of the wake is observed. These eddies are called twin vortices. When Re is over 90, eddies are continuously shed alternately from the two sides of the cylinder (Fig. 9.6(b)). Where $10^2 < Re < 10^5$, separation occurs near 80° from the front stagnation point (Fig. 9.6(c)). This arrangement of vortices is called a Kármán vortex street. Near $Re = 3.8 \times 10^5$, the boundary layer becomes turbulent and the separation position is moved further downstream to near 130° (Fig. 9.6(d)).

For a viscous fluid, as shown in Fig. 9.6, the flow lines along the cylinder surface separate from the cylinder to develop eddies behind it. This is visualised in Fig. 9.7. For the rear half of the cylinder, just like the case of a divergent pipe, the flow gradually decelerates with the velocity gradient reaching zero. This point is now the separation point, downstream of which flow reversals occur, developing eddies (see Section 7.4.2). This separation point shifts downstream as shown in Fig. 9.6(d) with increased $Re = Ud/v$ (d: cylinder diameter). The reason is that increased Re results in a turbulent boundary layer. Therefore, the fluid particles in and around the boundary layer mix with each other by the mixing action of the turbulent flow to make

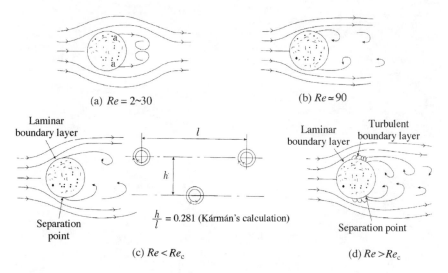

(a) $Re = 2\sim30$ (b) $Re \approx 90$

Laminar boundary layer

l

h

$\dfrac{h}{l} = 0.281$ (Kármán's calculation)

Separation point

(c) $Re < Re_c$

Laminar boundary layer   Turbulent boundary layer

Separation point

(d) $Re > Re_c$

Fig. 9.6 Flow around a cylinder

[1] Streeter, V.L., *Handbook of Fluid Dynamics*, (1961), McGraw-Hill, New York.

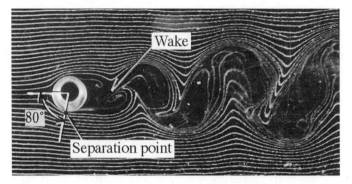

Fig. 9.7 Separation and Kármán vortex sheet (hydrogen bubble method) in water, velocity 2.4 cm/s, $Re = 195$

separation harder to occur. Figure 9.8 shows a flow visualisation of the development process from twin vortices to a Kármán vortex street. The Reynolds number $Re = 3.8 \times 10^5$ at which the boundary layer becomes turbulent is called the critical Reynolds number Re_c.

The pressure distribution on the cylinder surface in this case is like curves A, B and C in Fig. 9.5 with a reduced pressure behind the cylinder acting to produce a force in the downstream direction.

Figure 9.9 shows, for a cylinder of diameter d placed with its axis normal to a uniform flow U, changes in drag coefficient C_D with Re and also

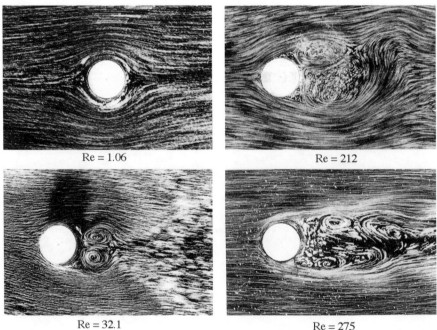

Re = 1.06   Re = 212

Re = 32.1   Re = 275

Fig. 9.8 Flow around a cylinder

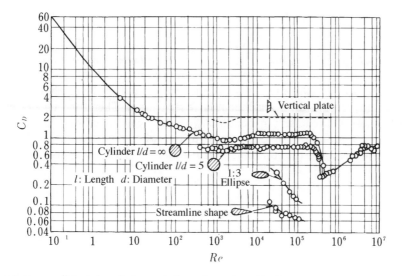

Fig. 9.9 Drag coefficients for cylinders and other column-shaped bodies

comparison with oblong and streamlined columns.[2] When $Re = 10^3 \sim 2 \times 10^5$, $C_D = 1 \sim 1.2$ or a roughly constant value; but when $Re = 3.8 \times 10^5$ or so, C_D suddenly decreases to 0.3. To explain this phenomenon, it is surmised that the location of the separation point suddenly changes as it reaches this Re, as shown in Fig 9.6(d).

G. I. Taylor (1886–1975, scholar of fluid dynamics at Cambridge University) calculated the number of vortices separating from the body every second, i.e. developing frequency f for $250 < Re < 2 \times 10^5$, by the following equation:

$$f = 0.198 \frac{U}{d}\left(1 - \frac{19.7}{Re}\right) \tag{9.7}$$

fd/U is a dimensionless parameter called the Strouhal number St (named after V. Strouhal (1850-1922), a Czech physicist; in 1878, he first investigated the 'singing' of wires), which can be used to indicate the degree of regularity in a cyclically fluctuating flow.

When the Kármán vortices develop, the body is acted on by a cyclic force and, as a result, it sometimes vibrates to produce sounds. The phenomenon where a power line 'sings' in the wind is an example of this.

In general, most drag is produced because a stream separates behind a body, develops vortices and lowers its pressure. Therefore, in order to reduce the drag, it suffices to make the body into a shape from which the flow does not separate. This is the so-called streamline shape.

---

[2] Hoerner, S.F., *Fluid Dynamic Drag*, (1965), Hoerner, Midland Park, NJ.

9.3.3 Drag of a sphere

The drag coefficient of a sphere changes as shown in Fig. 9.10.[3] Within the range where Re is fairly high, $Re = 10^3 \sim 2 \times 10^5$, the resistance is proportional to the square of the velocity, and C_D is approximately 0.44. As Re reaches 3×10^5 or so, like the case of a cylinder the boundary layer changes from laminar flow separation to turbulent flow separation. Therefore, C_D decreases to 0.1 or less. On reaching higher Re, C_D gradually approaches 0.2.

Slow flow around a sphere is known as Stokes flow. From the Navier–Stokes equation and the continuity equation the drag D is as follows:

$$\left. \begin{array}{r} D = 3\pi\mu U d \\ C_D = \dfrac{24}{Re} \end{array} \right\} \qquad (9.8)$$

This is known as Stokes' equation.[4] This coincides well with experiments within the range of $Re < 1$.

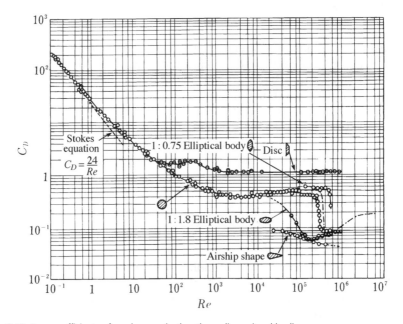

Fig. 9.10 Drag coefficients of a sphere and other three-dimensional bodies

9.3.4 Drag of a flat plate

As shown in Fig. 9.11, as a uniform flow of velocity U flows parallel to a flat plate of length l, the boundary layer steadily develops owing to viscosity.

[3] Streeter, V.L., *Handbook of Fluid Dynamics*, (1961), McGraw-Hill, New York.
[4] Lamb, H., *Hydrodynamics*, 6th edition, (1932), Cambridge University Press.

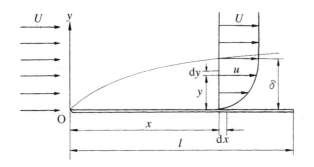

Fig. 9.11 Flow around a flat plate

Now, set the thickness of the boundary layer at a distance x from the leading edge of the flat plate to δ. Consider the mass flow rate of the fluid $\rho u\,dy$ flowing in the layer dy within the boundary layer at the given point x. From the difference in momentum of this flow quantity $\rho u\,dy$ before and after passing over this plate, the drag D due to the friction on the plate is as follows:

$$D = \int_0^\delta \rho u(U - u)\,dy \tag{9.9}$$

Now, putting the wall face friction stress as τ_0, and since $dD = \tau_0\,dx$, then from above

$$\tau_0 = \frac{dD}{dx} = \rho \frac{d}{dx}\int_0^\delta u(U - u)\,dy \tag{9.10}$$

Laminar boundary layer

Now, treating the distribution of u as a parabolic velocity distribution like the laminar flow in a circular pipe,

$$\eta = \frac{y}{\delta} \qquad \frac{u}{U} = 2\eta - \eta^2 \tag{9.11}$$

Substituting the above into eqn (9.10),

$$\tau_0 = \rho U^2 \frac{d\delta}{dx}\int_0^1 \frac{u}{U}\left(1 - \frac{u}{U}\right)d\eta = 0.133\rho U^2 \frac{d\delta}{dy} \tag{9.12}$$

On the other hand,

$$\tau_0 = \mu\left|\frac{du}{dy}\right|_{y=0} = 2\frac{\mu U}{\delta} \tag{9.13}$$

Therefore, from eqns (9.12) and (9.13),

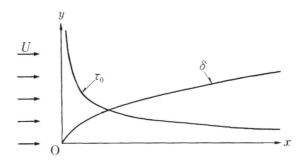

Fig. 9.12 Changes in boundary layer thickness and friction stress along a flat plate

$$\delta \, d\delta = 15.04 \frac{\mu}{\rho U} dx$$

$$\frac{\delta^2}{2} 15.04 \frac{v}{U} x + c$$

From $x = 0$ and $\delta = 0$, $c = 0$. Therefore

$$\delta = 5.48 \sqrt{\frac{vx}{U}} = \frac{5.48}{\sqrt{R_x}} x \tag{9.14}$$

However, since $R = U_x/v$, substitute eqn (9.14) into (9.13),

$$\tau_0 = 0.365 \sqrt{\frac{\mu \rho U^3}{x}} = 0.730 \frac{\rho U^2}{2} \sqrt{\frac{v}{Ux}} \tag{9.15}$$

As shown in Fig. 9.12, the boundary layer thickness δ increases in proportion to $\sqrt{x}$, while the surface frictional stress reduces in inverse proportion to $\sqrt{x}$.

The friction resistance for width b of the whole (but one face only) of that plate is expressed as follows by integrating eqn (9.15):

$$D = \int_0^l \tau_0 \, dx = 0.73 \sqrt{\mu \rho U^3 l} \, b \tag{9.16}$$

$$D = C_f l \frac{\rho U^2}{2} \tag{9.17}$$

Defining the friction drag coefficient as C_f, this becomes

$$C_f = \frac{1.46}{\sqrt{R_l}} \tag{9.18}$$

where $R = Ul/v$. The above equations roughly coincide with experimental values within the range of $R < 5 \times 10^5$.

Turbulent boundary layer
Whenever R_l is large, the length of laminar boundary layer is so short that the layer can be regarded as a turbulent boundary layer over the full length of a flat plate. Now, assume the distribution of u to be given by

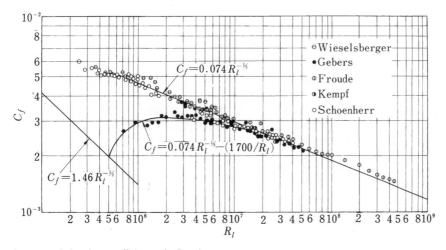

Fig. 9.13 Friction drag coefficients of a flat plate

$$\frac{u}{U} = \left(\frac{y}{\delta}\right)^{1/7} = \eta^{1/7} \tag{9.19}$$

like turbulent flow in a circular pipe, and the following equations are obtained:[5]

$$\delta = \frac{0.37x}{Rx^{1/5}} \tag{9.20}$$

$$\tau_0 = 0.029\rho U^2\left(\frac{v}{Ux}\right)^{1/5} \tag{9.21}$$

$$D = \frac{0.036\rho U^2 l}{R_l^{1/5}} \tag{9.22}$$

$$C_f = 0.072R_l^{-1/5} \tag{9.23}$$

The above equations coincide well with experimental values within the range of $5 \times 10^5 < R_l < 10^7$. From experimental data,

$$C_f = 0.074R_l^{-1/5} \tag{9.24}$$

gives better agreement.

In the case where there is a significant length of laminar boundary layer at the front end of a flat plate, but later developing into a turbulent boundary layer, eqn (9.12) is amended as follows:

$$C_f = \frac{0.074}{R_l^{1/5}} - \frac{1700}{R_l} \tag{9.25}$$

The relationship of C_f with R_l is shown in Fig. 9.13.

[5] Streeter, V. L., *Handbook of Fluid Dynamics*, (1961), McGraw-Hill, New York.

9.3.5 Friction torque acting on a revolving disc

If a disc revolves in a fluid at angular velocity ω, a boundary layer develops around the disc owing to the fluid viscosity.

Now, as shown in Fig. 9.14, let the radius of the disc be r_0, the thickness be b, and the resistance acting on the elementary ring area $2\pi r\, dr$ at a given radius $r\omega$ be dF. Assuming that dF is proportional to the square of the circular velocity $r\omega$ of that section, and the friction coefficient is f, the torque T_1 due to this surface friction is as follows:

$$dF = f\frac{\rho(r\omega)^2}{2}\, 2\pi r\, dr$$

or

$$T_1 = \int_{r=0}^{r_0} r\, dF = \frac{\pi f}{5}\rho\omega^2 r_0^5 \tag{9.26}$$

Now, putting the friction coefficient at the cylindrical part of the disc to f' and the resistance acting on it to F',

$$F' = f'\frac{\rho(r_0\omega)^2}{2}\, 2\pi r_0 b$$

Torque T_2 due to this surface friction is as follows:

$$T_2 = \pi f'\rho\omega^2 r_0^4 b \tag{9.27}$$

Assuming $f = f'$, the torque T needed for rotating this disc is

$$T = 2T_1 + T_2 = \pi f\rho\omega^2 r_0^4\left(\frac{2}{5}r_0 + b\right) \tag{9.28}$$

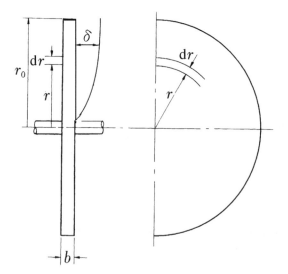

Fig. 9.14 A revolving disc

and the power L needed in that case is

$$L = T\omega = \pi f \rho \omega^3 r_0^4 \left(\frac{2}{5} r_0 + b \right)$$ (9.29)

These relationships are used for such cases as computing the power loss due to the friction of the impeller of a centrifugal pump or water turbine.

9.4 The lift of a body

9.4.1 Development of lift

Consider a case where, as shown in Fig. 9.15, a cylinder placed in a uniform flow U rotates at angular velocity ω but without flow separation. Since the fluid on the cylinder surface moves at a circular velocity $u = r_0 \omega$, sticking to the cylinder owing to the viscosity of the fluid, the flow velocity at a given point on the cylinder surface (angle θ) is the tangential velocity v_θ caused by the uniform flow U plus u. In other words, $2U \sin \theta + r_0 \omega$.

Putting the pressure of the uniform flow as p_∞, and the pressure at a given point on the cylinder surface as p, while neglecting the energy loss because it is too small, then from Bernoulli's equation

$$p_\infty + \frac{\rho}{2} U^2 = p + \frac{\rho}{2} (2U \sin \theta + r_0 \omega)^2$$

Therefore

$$\frac{p - p_\infty}{\rho U^2 / 2} = 1 - \left(\frac{2U \sin \theta + r_0 \omega}{U} \right)^2$$ (9.30)

Consequently, for unit width of the cylinder surface, integrate the component in the y direction of the force due to the pressure $p - p_\infty$ acting on a minute area $r_0 \, d\theta$, and the lift L acting on the unit width of cylinder is obtained:

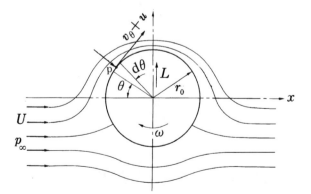

Fig. 9.15 Lift acting on a rotating cylinder

$$L = 2 \int_{-\pi/2}^{\pi/2} -(p - p_\infty) r_0 \, d\theta \sin\theta$$

$$= -r_0 \rho U^2 \int_{-\pi/2}^{\pi/2} \left[1 - \left(\frac{2U \sin\theta + r_0\omega}{U} \right)^2 \right] \sin\theta \, d\theta$$

$$= -r_0 \rho U^2 \int_{-\pi/2}^{\pi/2} \left[1 - \left(\frac{r_0\omega}{U} \right)^2 - \frac{4r_0\omega}{U} \sin\theta - 4\sin^2\theta \right] \sin\theta \, d\theta$$

$$= 2\pi r_0^2 \omega \rho U = 2\pi r_0 u \rho U \tag{9.31}$$

The circulation around the cylinder surface when a cylinder placed in a uniform flow U has circular velocity u is

$$\Gamma = 2\pi r_0 u$$

Substituting the above into eqn (9.31),

$$L = \rho U \Gamma \tag{9.32}$$

This lift is the reason why a baseball, tennis ball or golf ball curves or slices if spinning.[6] This equation is called the Kutta–Joukowski equation.

In general, whenever circulation develops owing to the shape of a body placed in the uniform flow U (e.g. aircraft wings or yacht sails) (see Section 9.4.2), lift L as in eqn (9.32) is likewise produced for the unit width of its section.

9.4.2 Wing

Of the forces acting on a body placed in a flow, if the body is so manufactured as to make the lift larger than the drag, it is called a wing, aerofoil or blade.

The shape of a wing section is called an aerofoil section, an example of which is shown in Fig. 9.16. The line connecting the leading edge with the trailing edge is called the chord, and its length is called the chord length. The line connecting the mid-points of the upper and lower faces of the aerofoil

[6] The reason why the golf ball surface has many dimple-like hollows is to reduce the air resistance by producing turbulence around the ball, and to produce an effective lift while keeping a stable flight by making the air circulation larger (see Plate 4). The number of rotations (called spin) per second of a golf ball can be 100 or more.

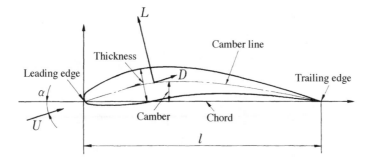

Fig. 9.16 An aerofoil section

section is called the camber line. The height of the camber line from the chord is called the camber, which mostly means its maximum value in particular. The thickness of a wing as measured vertically to the camber or chord is called its thickness, whose maximum value is called the maximum thickness. Furthermore, the angle α between the chord and the flow direction U is called the angle of attack. Putting the wing width as b, and the maximum projection area of the wing as A, b^2/A is called the aspect ratio. Assuming the length of the chord is l, since $A = bl$ for an oblong wing, the aspect ratio becomes $b^2/A = b/l$.

Of the studies on the characteristics of wing shape, those most well known have been performed by the USA's NACA (National Advisory Committee for Aeronautics, renamed in 1959 as NASA (National Aeronautics and Space Administration)), by the UK's RAE (Royal Aircraft Establishment) and by Germany's Göttingen University. The particular wing shapes are named after them.

The lift L, drag D and moment M (moment about the wing leading edge or the point on the chord $l/4$ from the leading edge) acting on the wing are expressed respectively for unit width by the following equations:

$$\left. \begin{array}{l} L = C_L l \dfrac{\rho U^2}{2} \\[2mm] D = C_D l \dfrac{\rho U^2}{2} \\[2mm] M = C_M l^2 \dfrac{\rho U^2}{2} \end{array} \right\} \tag{9.33}$$

C_L, C_D and C_M are called respectively the lift coefficient, drag coefficient and moment coefficient to be determined by the aerofoil section, Mach number and Reynolds number. The wing characteristic is indicated by the values of C_L, C_D and C_M for the angle of attack α, or by plotting C_D and C_M on the abscissa and C_L on the ordinate. These plots are called the characteristic curves. Some examples of them are shown in Figs 9.17, 9.19 and 9.20.

The lift coefficient C_L reaches zero at a certain angle of attack α, called the zero lift angle. As the angle of attack increases from the zero lift angle,

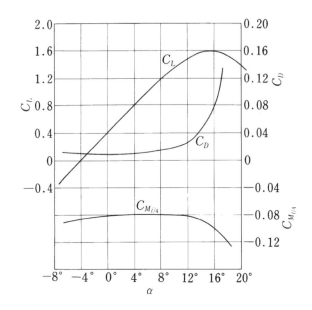

Fig. 9.17 Characteristic curves of a wing

the lift coefficient C_L increases in a straight line. As it further increases, however, the increase in C_L gradually slows down, reaches a maximum value at a certain point, and thereafter suddenly decreases. This is due to the fact that, as shown in Fig. 9.18, the flow separates on the upper surface of the wing because the angle of attack has increased too much. This phenomenon is completely analogous to the separation occurring on a divergent pipe or flow behind a body and is called stall. Angle α at which C_L reaches a maximum is the stalling angle and the maximum value of C_L is the maximum lift coefficient. Figure 9.19 shows the characteristic with changing wing section.

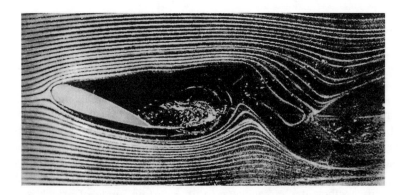

Fig. 9.18 Flow around a stalled wing

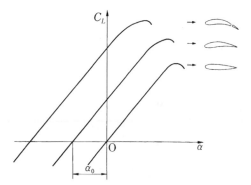

Fig. 9.19 Aerofoil section and characteristic

Figure 9.20 shows a wing characteristic by putting C_D on the abscissa and C_L on the ordinate, and is called the lift–drag polar, from which the angle of attack maximising the lift–drag ratio C_L/C_D can easily be found.

The reason why a wing produces lift is because a circulatory flow is produced just like for a rotating cylinder. In the case of a wing section, the circulatory flow is produced because the trailing edge is sharpened. A wing moves from a stationary state initially as shown in Fig. 9.21(a). Owing to its behaviour as potential flow, a rear stagnation point develops at point A. Consequently, the flow develops into a flow running round the trailing edge B. Since the trailing edge is sharp, however, the flow is unable to run round the wing surface but separates from it producing a vortex as shown in (b) of

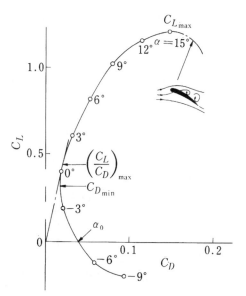

Fig. 9.20 Characteristic curve of a wing (lift–drag curve)

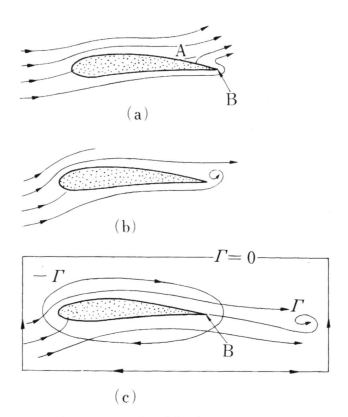

Fig. 9.21 Development of circulation around aerofoil section

the same figure. This vortex moves backwards being driven by the main flow. The flow on the upper surface of the wing is drawn towards the trailing edge, which itself develops into a stagnation point, and thus the flow is now as shown in (c) of the same figure. As one vortex is produced, another vortex of equal strength is also produced since the flow system as a whole should be in a net non-rotary movement. Therefore a circulation is produced against the start-up vortex as if another vortex of equal strength in counterrotation had developed around the wing section. The former vortex is called a starting-up vortex because it is left at the starting point; the latter assumed vortex is a wing-bound vortex. The situation where the flow runs off the sharp trailing edge of a wing as stated above is called the Kutta condition or Joukowski's hypothesis. Figure 9.22 shows the visualised picture of a starting vortex.

The blades of a blower, compressor, water wheel, steam turbine or gas turbine of the axial flow type are distributed radially in planes around the shaft and the blade sections of the same shape are found arranged at a certain spacing as shown in Fig. 9.23. This is called a cascade.

The action of a cascade is to change the flow direction with small loss by using the necessary stagger angle.

The lift acting on a blade is expressed by $\rho v_\infty \Gamma$ from eqn (9.32) where v_∞

Fig. 9.22 Starting vortex (courtesy of the National Physical Laboratory)

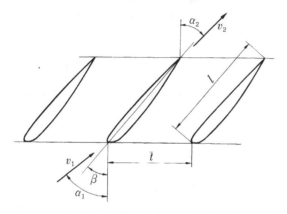

Fig. 9.23 Cascade: v_1, v_2, velocities at infinity in front and behind the cascade; α_1, inlet angle (angle of velocity v_1 to axial direction); α_2, exit angle (angle of velocity v_2 to axial direction); l, chord length; t, space between blades; l/t, solidity; β, stagger angle; $\theta = \alpha_1 - \alpha_2$, turning angle of flow

represents the mean flow velocity of v_1 and v_2. The magnitude of the circulation around a blade in a cascade is affected by the other blades giving less lift compared with a solitary blade.

For the same blade section, setting the lifts of a solitary blade and a cascade blade to L_0 and L respectively,

$$k = L/L_0 \tag{9.34}$$

k is called the interference coefficient. It is a function of l/t and β, and is near one whenever l/t is 0.5 or less.

9.5 Cavitation

According to Bernoulli's principle, as the velocity increases, the pressure decreases correspondingly. In the forward part on the upper surface of a wing

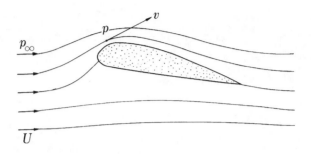

Fig. 9.24 An aerofoil section inside the flow

section placed in a uniform flow as shown in Fig. 9.24, for example, the flow velocity increases while the pressure decreases.

If a section of a body placed in liquid increases its velocity so much that the pressure there is less than the saturation pressure of the liquid, the liquid instantaneously boils, producing bubbles with cavities. This phenomenon is called cavitation. In addition, since gas dissolves in liquid in proportion to the pressure (Henry's Law), as the liquid pressure decreases, the dissolved gas separates from the liquid into bubbles even before the saturation pressure is reached. When these bubbles are conveyed downstream where the pressure is higher they are suddenly squeezed and abnormally high pressure develops.[7] At this point noise and vibration occur eroding the neighbouring surface and leaving on it holes small in diameter but relatively deep, as if made by a slender drill in most cases. These phenomena as a whole are also referred to as cavitation in a wider sense.

The blades of a pump or water wheel, or the propeller of a boat, are sometimes destroyed by such phenomena. They can develop on liquid-carrying pipe lines or on hydraulic devices and cause failures.

The saturation pressures at various temperatures are shown in Table 9.3, while the volume ratios of air soluble in water at 1 atm are given in Table 9.4.

When an aerofoil section is placed in a flow of liquid, the pressure distribution on its surface is as shown in Fig. 9.25. As the cavity grows, the upper pressure characteristic curve lowers while vibration etc. grow. When the liquid pressure is low and the flow velocity is large, the cavity grows further. When it grows beyond twice the chord length, the flow stabilises, with noise and vibration reducing. This situation is called supercavitation, and is applied to the wings of a hydrofoil boat.

Let the upstream pressure not affected by the wing be p_∞, the flow velocity U and the saturation pressure p_v. When the pressure at a point on the wing surface or nearby has reached p_v, cavitation develops. The ratio of $p_\infty - p_v$ to the dynamic pressure is expressed by the following equation:

---

[7] According to actual measurements, a pressure of 100–200 atmospheres, or sometimes as high as 500 atmospheres, is brought about.

Table 9.3 Saturation pressure for water

Temp. (°C)	Pa	Temp. (°C)	Pa
0	608	50	12 330
10	1226	60	19 920
20	2334	70	31 160
30	4236	80	47 360
40	7375	100	101 320

Table 9.4 Solubility of air in water

Temp. (°C)	0	20	40	60	80	100
Air	0.028 8	0.018 7	0.014 2	0.012 2	0.011 3	0.011 1

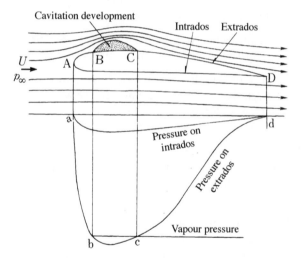

Fig. 9.25 Development of cavitation on an aerofoil section

$$k_d = \frac{p_\infty - p_v}{\rho U^2 / 2} \tag{9.35}$$

k_d in this equation is called the cavitation number. When k_d is small, cavitation is likely to develop.

9.6 Problems

1. Obtain the terminal velocity of a spherical sand particle dropping freely in water.

2. A wind of velocity 40 m/s is blowing against an electricity pole 50 cm in diameter and 5 m high. Obtain the drag and the maximum bending moment acting on the pole. Assume that the drag coefficient is 0.6 and the air density is 1205 kg/m^3.

3. A smooth spherical body of diameter 12 cm is travelling at a velocity of 30 m/s in windless open air under the conditions of 20°C temperature and standard atmospheric pressure. Obtain the drag of the sphere.

4. If air at standard atmospheric pressure is flowing at velocity of 4 km/h along a flat plate of length 2.5 m, what is the maximum value of the boundary layer thickness? What is it when the wind velocity is 120 km/h?

5. What are the torque and the power necessary to turn a rotor as shown in Fig. 9.13 at 600 rpm in oil of specific gravity 0.9? Assume that the friction coefficient $f = 0.147$, $r_0 = 30$ cm and $b = 5$ cm.

6. When walking on a country road in a cold wintry wind, whistling sounds can be heard from power lines blown by the wind. Explain the phenomenon by which such sounds develop.

7. In a baseball game, when the pitcher throws a drop or a curve, the ball significantly and suddenly goes down or curves. Find out why.

8. An oblong barge of length 10 m, width 2.5 m and draft 0.25 m is going up a river at a relative velocity of 1.5 m/s to the water flow. What are the friction resistance suffered by the barge and the power necessary for navigation, assuming a water temperature of 20°C?

9. If a cylinder of radius $r = 3$ cm and length $l = 50$ cm is rotating at $n = 1000$ rpm in air where a wind velocity $u = 10$ m/s, how much lift is produced on the cylinder? Assume that $\rho = 1.205$ kg/m^3 and that air on the cylinder surface does not separate.

10. A car of frontal projection area 2 m^2 is running at 60 km/h in calm air of temperature 20°C and standard atmospheric pressure. What is the drag on the car? Assume that the resistance coefficient is 0.4.

10

Dimensional analysis and law of similarity

The method of dimensional analysis is used in every field of engineering, especially in such fields as fluid dynamics and thermodynamics where problems with many variables are handled. This method derives from the condition that each term summed in an equation depicting a physical relationship must have same dimension. By constructing non-dimensional quantities expressing the relationship among the variables, it is possible to summarise the experimental results and to determine their functional relationship.

Next, in order to determine the characteristics of a full-scale device through model tests, besides geometrical similarity, similarity of dynamical conditions between the two is also necessary. When the above dimensional analysis is employed, if the appropriate non-dimensional quantities such as Reynolds number and Froude number are the same for both devices, the results of the model device tests are applicable to the full-scale device.

10.1 Dimensional analysis

When the dimensions of all terms of an equation are equal the equation is dimensionally correct. In this case, whatever unit system is used, that equation holds its physical meaning. If the dimensions of all terms of an equation are not equal, dimensions must be hidden in coefficients, so only the designated units can be used. Such an equation would be void of physical interpretation.

Utilising this principle that the terms of physically meaningful equations have equal dimensions, the method of obtaining dimensionless groups of which the physical phenomenon is a function is called dimensional analysis.

If a phenomenon is too complicated to derive a formula describing it, dimensional analysis can be employed to identify groups of variables which would appear in such a formula. By supplementing this knowledge with experimental data, an analytic relationship between the groups can be constructed allowing numerical calculations to be conducted.

10.2 Buckingham's π theorem

In order to perform the dimensional analysis, it is convenient to use the π theorem. Consider a physical phenomenon having n physical variables v_1, v_2, $v_3, \ldots, v_n$ and k basic dimensions[1] (L, M, T or L, F, T or such) used to describe them. The phenomenon can be expressed by the relationship among $n - k = m$ non-dimensional groups $\pi_1, \pi_2, \pi_3, \ldots, \pi_m$. In other words, the equation expressing the phenomenon as a function f of the physical variables

$$f(v_1, v_2, v_3, \ldots, v_n) = 0 \tag{10.1}$$

can be substituted by the following equation expressing it as a function ϕ of a smaller number of non-dimensional groups:

$$\phi(\pi_1, \pi_2, \pi_3, \ldots, \pi_m) = 0 \tag{10.2}$$

This is called Buckingham's π theorem. In order to produce $\pi_1, \pi_2, \pi_3, \ldots, \pi_m$, k core physical variables are selected which do not form a π themselves. Each π group will be a power product of these with each one of the m remaining variables. The powers of the physical variables in each π group are determined algebraically by the condition that the powers of each basic dimension must sum to zero.

By this means the non-dimensional quantities are found among which there is the functional relationship expressed by eqn (10.2). If the experimental results are arranged in these non-dimensional groups, this functional relationship can clearly be appreciated.

10.3 Application examples of dimensional analysis

10.3.1 Flow resistance of a sphere

Let us study the resistance of a sphere placed in a uniform flow as shown in Fig. 10.1. In this case the effect of gravitational and buoyancy forces will be neglected. First of all, as the physical quantities influencing the drag D of a sphere, sphere diameter d, flow velocity U, fluid density ρ and fluid viscosity μ, are candidates. In this case $n = 5$, $k = 3$ and $m = 5 - 3 = 2$, so the number of necessary non-dimensional groups is two. Select ρ, U and d as the k core physical quantities, and the first non-dimensional group π, formed with D, is

$$\pi_1 = D\rho^x U^y d^z = [LMT^{-2}][L^{-3}M]^x[LT^{-1}]^y[L]^z$$
$$= L^{1-3x+y+z}M^{1+x}T^{-2-y} \tag{10.3}$$

---

[1] In general the basic dimensions in dynamics are three – length [L], mass [M] and time [T] – but as the areas of study, e.g. heat and electricity, expand, the number of basic dimensions increases.

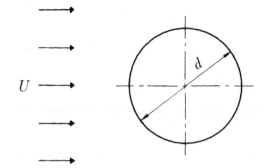

Fig. 10.1 Sphere in uniform flow

i.e.

$$L: \quad 1 - 3x + y + z = 0$$
$$M: \quad 1 + x = 0$$
$$T: \quad -2 - y = 0$$

Solving the above simultaneously gives

$$x = -1 \qquad y = -2 \qquad z = -2$$

Substituting these values into eqn (10.3), then

$$\pi_1 = \frac{D}{\rho U^2 d^2} \tag{10.4}$$

Next, select μ with the three core physical variables in another group, and

$$\pi_2 = \mu \rho^x U^y d^z = [L^{-1} M T^{-1}][L^{-3} M]^x [L T^{-1}]^y [L]^z$$
$$= L^{1-3x+y+z} M^{1+x} T^{-1-y} \tag{10.5}$$

i.e.

$$L: \quad -1 - 3x + y + z = 0$$
$$M: \quad 1 + x = 0$$
$$T: \quad -1 - y = 0$$

Solving the above simultaneously gives

$$x = -1 \qquad y = -1 \qquad z = -1$$

Substituting these values into eqn (10.5), then

$$\pi_2 = \frac{D}{\rho U d} \tag{10.6}$$

Therefore, from the π theorem the following functional relationship is obtained:

$$\pi_1 = f(\pi_2) \tag{10.7}$$

Consequently

$$\frac{D}{\rho U^2 d^2} = f\left(\frac{D}{\rho U d}\right) \qquad (10.8)$$

In eqn (10.8), since d^2 is proportional to the projected area of sphere $A = (\pi d^2/4)$, and $\rho U d/\mu = U d/\nu = Re$ (Reynolds number), the following general expression is obtained:

$$D = C_D A \frac{\rho U^2}{2} \qquad (10.9)$$

where $C_D = f(Re)$. Equation (10.9) is just the same as eqn (9.4). Since C_D is found to be dependent on Re, it can be obtained through experiment and plotted against Re. The relationship is that shown in Fig. 9.10. Even through this result is obtained through an experiment using, say, water, it can be applied to other fluids such as air or oil, and also used irrespective of the size of the sphere. Furthermore, the form of eqn (10.9) is always applicable, not only to the case of the sphere but also where the resistance of any body is studied.

10.3.2 Pressure loss due to pipe friction

As the quantities influencing pressure loss $\Delta p/l$ per unit length due to pipe friction, flow velocity v, pipe diameter d, fluid density ρ, fluid viscosity μ and pipe wall roughness ε, are candidates. In this case, $n = 6$, $k = 3$, $m = 6 - 3 = 3$.

Obtain π_1, π_2, π_3 by the same method as in the previous case, with ρ, v and d as core variables:

$$\pi_1 = \frac{\Delta p}{l} \rho^x v^y d^z = [L^{-3}F][L^{-4}FT^2]^x[LT^{-1}]^y[L]^z = \frac{\Delta p}{l} \frac{d}{\rho v^2} \qquad (10.10)$$

$$\pi_2 = \mu \rho^x v^y d^z = [L^{-2}FT][L^{-4}FT^2]^x[LT^{-1}]^y[L]^z = \frac{\mu}{\rho v d} \qquad (10.11)$$

$$\pi_3 = \varepsilon \rho^x v^y d^z = [L][L^{-4}FT^2]^x[LT^{-1}]^y[L]^z = \frac{\varepsilon}{d} \qquad (10.12)$$

Therefore, from the π theorem, the following functional relationship is obtained:

$$\pi_1 = f(\pi_2, \pi_3) \qquad (10.13)$$

and

$$\frac{\Delta p}{l} \frac{d}{\rho v^2} = f\left(\frac{\mu}{\rho v d}, \frac{\varepsilon}{d}\right)$$

That is,

$$\Delta p = \frac{l}{d} \rho v^2 f\left(\frac{\mu}{\rho v d}, \frac{\varepsilon}{d}\right) \qquad (10.14)$$

The loss of head h is as follows:

$$h = \frac{\Delta p}{\rho g} = f\left(\frac{1}{Re}, \frac{\varepsilon}{d}\right)\frac{l}{d}\frac{v^2}{2g} = \lambda \frac{l}{d}\frac{v^2}{2g} \tag{10.15}$$

where $\lambda = f(Re, \varepsilon/d)$. Equation (10.15) is just the same as eqn (7.4), and λ can be summarised against Re and ε/d as shown in Figs 7.4 and 7.5.

10.4 Law of similarity

When the characteristics of a water wheel, pump, boat or aircraft are obtained by means of a model, unless the flow conditions are similar in addition to the shape, the characteristics of the prototype cannot be assumed from the model test result. In order to make the flow conditions similar, the respective ratios of the corresponding forces acting on the prototype and the model should be equal. The forces acting on the flow element are due to gravity F_G, pressure F_P, viscosity F_V, surface tension F_T (when the prototype model is on the boundary of water and air), inertia F_I and elasticity F_E.

The forces can be expressed as shown below.

gravity force	$F_G = mg = \rho L^3 g$
pressure force	$F_P = (\Delta p)A = (\Delta p)L^2$
viscous force	$F_V = \mu\left(\frac{du}{dy}\right)A = \mu\left(\frac{v}{L}\right)L^2 = \mu v L$
surface tension force	$F_T = TL$
inertial force	$F_I = ma = \rho L^3 \dfrac{L}{T^2} = \rho L^4 T^{-2} = \rho v^2 L^2$
elasticity force	$F_E = KL^2$

Since it is not feasible to have the ratios of all such corresponding forces simultaneously equal, it will suffice to identify those forces that are closely related to the respective flows and to have them equal. In this way, the relationship which gives the conditions under which the flow is similar to the actual flow in the course of a model test is called the law of similarity. In the following section, the more common force ratios which ensure the flow similarity under appropriate conditions are developed.

10.4.1 Non-dimensional groups which determine flow similarity

Reynolds number
Where the compressibility of the fluid may be neglected and in the absence of a free surface, e.g. where fluid is flowing in a pipe, an airship is flying in the air (Fig. 10.2) or a submarine is navigating under water, only the viscous force and inertia force are of importance:

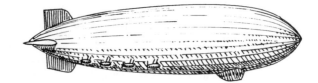

Fig. 10.2 Airship

$$\frac{\text{inertia force}}{\text{viscous force}} = \frac{F_I}{F_V} = \frac{\rho v^2 L^2}{\mu v L} = \frac{Lv\rho}{\mu} = \frac{Lv}{\nu} = Re$$

which defines the Reynolds number Re,

$$Re = Lv/\nu \qquad (10.16)$$

Consequently, when the Reynolds numbers of the prototype and the model are equal the flow conditions are similar. Equations (10.16) and (4.5) are identical.

Froude number

When the resistance due to the waves produced by motion of a boat (gravity wave) is studied, the ratio of inertia force to gravity force is important:

$$\frac{\text{inertia force}}{\text{gravity force}} = \frac{F_I}{F_G} = \frac{\rho v^2 L^2}{\rho L^3 g} = \frac{v^2}{gL}$$

In general, in order to change v^2 above to v as in the case for Re, the square root of v^2/gL is used. This square root is defined as the Froude number Fr,

$$Fr = \frac{v}{\sqrt{gL}} \qquad (10.17)$$

If a test is performed by making the Fr of the actual boat (Fig. 10.3) and of the model ship equal, the result is applicable to the actual boat so far as the wave resistance alone is concerned. This relationship is called Froude's law of similarity. For the total resistance, the frictional resistance must be taken into account in addition to the wave resistance.

Also included in the circumstances where gravity inertia forces are

Fig. 10.3 Ship

important are flow in an open ditch, the force of water acting on a bridge pier, and flow running out of a water gate.

Weber number

When a moving liquid has its face in contact with another fluid or a solid, the inertia and surface tension forces are important:

$$\frac{\text{inertia force}}{\text{surface tension}} = \frac{F_I}{F_T} = \frac{\rho v^2 L^2}{TL} = \frac{\rho v^2 L}{T}$$

In this case, also, the square root is selected to be defined as the Weber number We,

$$We = v\sqrt{\rho L / T} \qquad (10.18)$$

We is applicable to the development of surface tension waves and to a poured liquid.

Mach number

When a fluid flows at high velocity, or when a solid moves at high velocity in a fluid at rest, the compressibility of the fluid can dominate so that the ratio of the inertia force to the elasticity force is then important (Fig. 10.4):

$$\frac{\text{inertia force}}{\text{elastic force}} = \frac{F_I}{F_E} = \frac{\rho v^2 L^2}{KL} = \frac{v^2}{K/\rho} = \frac{v^2}{a^2}$$

Again, in this case, the square root is selected to be defined as the Mach number M,

$$M = v/a \qquad (10.19)$$

$M < 1$, $M = 1$ and $M > 1$ are respectively called subsonic flow, sonic flow and supersonic flow. When $M = 1$ and $M < 1$ and $M > 1$ zones are coexistent, the flow is called transonic flow.

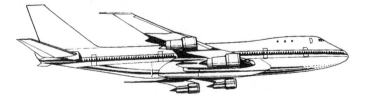

Fig. 10.4 Boeing 747: full length, 70.5 m; full width, 59.6 m; passenger capacity, 498 persons; turbofan engine and cruising speed of 891 km/h ($M = 0.82$)

10.4.2 Model testing

From such external flows as over cars, trains, aircraft, boats, high-rise buildings and bridges to such internal flows as in tunnels and various machines like pumps, water wheels, etc., the prediction of characteristics

Ernst Mach (1838–1916)
Austrian physicist/philosopher. After being professor at Graz and Prague Universities became professor at Vienna University. Studied high-velocity flow of air and introduced the concept of Mach number. Criticised Newtonian dynamics and took initiatives on the theory of relativity. Also made significant achievements in thermodynamics and optical science.

through model testing is widely employed. Suppose that the drag D on a car is going to be measured on a 1:10 model (scale ratio $S = 10$). Assume that the full length l of the car is 3 m and the running speed v is 60 km/h. In this case, the following three methods are conceivable. Subscript m refers to the model.

Test in a wind tunnel   In order to make the Reynolds numbers equal, the velocity should be $v_m = 167$ m/s, but the Mach number is 0.49 including compressibility. Assuming that the maximum tolerable value M of incompressibility is 0.3, $v_m = 102$ m/s and $Re_m/Re = v_m/Sv = 0.61$. In this case, since the flows on both the car and model are turbulent, the difference in C_D due to the Reynolds numbers is modest. Assuming the drag coefficients for both $D/(\rho v^2 l^2/2)$ are equal, then the drag is obtainable from the following equation:

$$D = D_m \left(\frac{v}{v_m} \frac{l}{l_m}\right)^2 = D_m \left(\frac{Sv}{v_m}\right)^2 \qquad (10.20)$$

This method is widely used.

Test in a circulating flume or towing tank   In order to make the Reynolds numbers for the car and the model equal, $v_m = vSv_m/v = 11.1$ m/s. If water is made to flow at this velocity, or the model is moved under calm water at this velocity, conditions of dynamical similarity can be realised. The conversion formula is

$$D = D_m \frac{\rho}{\rho_m} \left(\frac{Sv}{v_m}\right)^2 = D_m \frac{\rho v^2}{\rho_m v_m^2} \qquad (10.21)$$

Test in a variable density wind tunnel If the density is increased, the Reynolds numbers can be equalised without increasing the air flow velocity. Assume that the test is made at the same velocity; it is then necessary to increase the wind tunnel pressure to 10 atm assuming the temperatures are equal. The conversion formula is

$$D = D_\mathrm{m} \frac{\rho}{\rho_\mathrm{m}} S^2 \tag{10.22}$$

Two mysteries solved by Mach

[No. 1] The early Artillerymen knew that two bangs could be heard downrange from a gun when a high-speed projectile was fired, but only one from a low-speed projectile. But they did not know the reason and were mystified by these phenomena. Following Mach's research, it was realised that in addition to the bang from the muzzle of the gun, an observer downrange would first hear the arrival of the bow shock which was generated from the head of the projectile when its speed exceeded the velocity of sound.

By this reasoning, this mystery was solved.

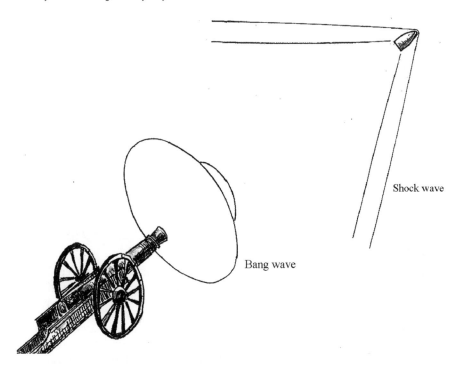

Shock wave

Bang wave

[No. 2] This is a story of the Franco-Prussian war of 1870–71. It was found that the novel French Chassepôt high-speed bullets caused large crater-shaped wounds. The French were suspected of using explosive projectiles and therefore violating the International Treaty of Petersburg prohibiting the use of explosive projectiles. Mach then gave the complete and correct explanation that the explosive type wounds were caused by the highly pressurised air caused by the bullet's bow wave and the bullet itself.

So it was clear that the French did not use explosive projectiles and the mystery was solved.

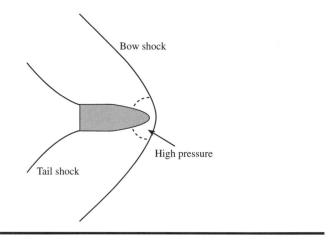

10.5 Problems

1. Derive Torricelli's principle by dimensional analysis.

2. Obtain the drag on a sphere of diameter d placed in a slow flow of velocity U.

3. Assuming that the travelling velocity a of a pressure wave in liquid depends upon the density ρ and the bulk modulus k of the liquid, derive a relationship for a by dimensional analysis.

4. Assuming that the wave resistance D of a boat is determined by the velocity v of the boat, the density ρ of fluid and the acceleration of gravity g, derive the relationship between them by dimensional analysis.

5. When fluid of viscosity μ is flowing in a laminar state in a circular pipe of length l and diameter d with a pressure drop Δp, obtain by dimensional analysis a relationship between the discharge Q and d, $\Delta p/l$ and μ.

6. Obtain by dimensional analysis the thickness δ of the boundary layer distance x along a flat plane placed in a uniform flow of velocity U (density ρ, viscosity μ).

7. Fluid of density ρ and viscosity μ is flowing through an orifice of diameter d bringing about a pressure difference Δp. For discharge Q, the

discharge coefficient $C = Q/[(\pi d^2/4)\sqrt{2\Delta p/\rho}]$, and $Re = d\sqrt{2\rho\Delta p}/\mu$, show by dimensional analysis that there is a relationship $C = f(Re)$.

8. An aircraft wing, chord length 1.2 m, is moving through calm air at 20°C and 1 bar at a velocity of 200 km/h. If a model wing of scale 1:3 is placed in a wind tunnel, assuming that the dynamical similarity conditions are satisfied by Re, then:

   (a) If the temperature and the pressure in the wind tunnel are respectively equal to the above, what is the correct wind velocity in the tunnel?
   (b) If the air temperature in the tunnel is the same but the pressure is increased by five times, what is the correct wind velocity? Assume that the viscosity μ is constant.
   (c) If the model is tested in a water tank of the same temperature, what is the correct velocity of the model?

9. Obtain the Froude number when a container ship of length 245 m is sailing at 28 knots. Also, when a model of scale 1:25 is tested under similarity conditions where the Froude numbers are equal, what is the proper towing velocity for the model in the water tank? Take 1 knot $= 0.514$ m/s.

10. For a pump of head H, representative size l and discharge Q, assume that the following similarity rule is appropriate:

$$\frac{l}{l_m} = \left(\frac{Q}{Q_m}\right)^{1/2}\left(\frac{H}{H_m}\right)^{-1/4}$$

where, for the model, subscript m is used.

   If a pump of $Q = 0.1\,\text{m}^3/\text{s}$ and $H = 40\,\text{m}$ is model tested using this relationship in the situation $Q_m = 0.02\,\text{m}^3/\text{s}$ and $H_m = 50\,\text{m}$, what is the model scale necessary for dynamical similarity?

11

Measurement of flow velocity and flow rate

To clarify fluid phenomena, it is necessary to measure such quantities as pressure, flow velocity and flow rate. Since the measurement of pressure was covered in Section 3.1.5, in this chapter we cover the measurement of flow velocity and flow rate. Fluid includes both gas and liquid. According to the type and condition of the fluid, or if it flows in a pipe line or open channel, various methods of measurement were developed and are in practical use.

11.1 Measurement of flow velocity

11.1.1 Pitot tube

Figure 11.1 shows the shape of a commonly used standard Pitot tube (also called a Pitot-static tube). The flow velocity is given by the following equation from total pressure p_1 and static pressure p_2, both to be measured as in the case of eqn (5.20):

$$v = c\sqrt{2(p_1 - p_2)/\rho} \qquad (11.1)$$

where c is called the Pitot tube coefficient, which may be taken as having value 1 for a standard-type Pitot tube. However, when compressibility is to be taken into account, refer to Section 13.4.

A Pitot tube is also used to measure the flow in a large-diameter pipe. In this case, the cross-section of the pipe is divided into ring-like equal areas, and the flow velocity at the centre of the area of every ring is measured. The mean flow velocity is obtained from their mean value, and the total flow rate is obtained from the product of the mean velocity and the section area. Apart from the standard type, there are various other types of Pitot tube, as follows.

Cylinder-type Pitot tube
This type of Pitot tube is used to measure simultaneously the direction and the flow velocity of a two-dimensional flow utilising the pressure distribution

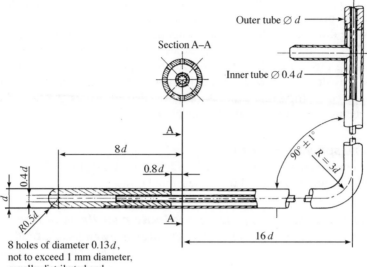

Section A–A

Outer tube ⌀ d

Inner tube ⌀ $0.4\,d$

$90° \pm 1°$ $R = 3d$

$8d$

$0.8\,d$

$0.4\,d$

d

$R0.5d$

$16\,d$

8 holes of diameter $0.13\,d$,
not to exceed 1 mm diameter,
equally distributed and
free from burrs

Fig. 11.1 NPL-type Pitot tube

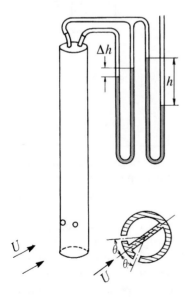

Δh

h

U

U

θ θ

U

Fig. 11.2 Cylinder-type Pitot tube

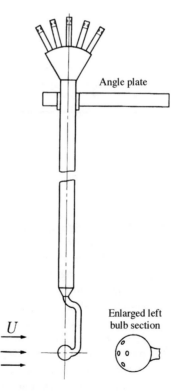

Angle plate

U

Enlarged left
bulb section

Fig. 11.3 Five-hole spherical Pitot tube

on the cylinder surface wall that is shown in Fig. 9.5. Figure 11.2 shows the measuring principle. The body is rotated in a flow until $\Delta h = 0$, and the centre-line direction is then the flow direction. The static pressure is obtained if $\theta = 33°\text{--}35°$. Then, if one of the holes is made to face the flow direction by rotating the cylinder, it measures the total pressure. If a third measuring hole is provided on the centre line, the flow direction and both pressures can be measured at the same time. A device which measures the flow direction and velocity in this way is called a yawmeter.

Five-hole spherical Pitot tube
This is constructed as shown in Fig. 11.3, and is capable of measuring the velocity and direction of a three-dimensional flow.

Pitot tube for measuring the flow velocity near the wall face
For measuring the velocity of a flow very near the wall face, a total pressure tube from a flattened fine tube as shown in Fig. 11.4(a) is used. For measuring the velocity of a flow even nearer to the wall face, a surface Pitot tube as shown in Fig. 11.4(b) is used. By changing the width of opening B while moving the tube, the whole pressure distribution can be determined. In this case, the static pressure is measured by another hole on the wall face.

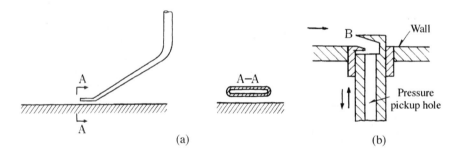

(a) (b)

Fig. 11.4 Pitot tubes for measuring the velocity of flow near the wall face: (a) total pressure tube; (b) surface Pitot tube

11.1.2 Hot-wire anemometer

If a heated fine wire is placed in a flow, the temperature of the hot wire changes according to the velocity of the fluid so changing its electrical resistance. A meter which measures the flow by utilising this change in resistance is called a hot-wire anemometer.

One method is shown in Fig. 11.5(a). The flow velocity is obtained by reading the changing hot-wire temperature as a change of electrical resistance (using the galvanometer G) while keeping the voltage between C and D constant. This is called the constant voltage anemometer. A second method is shown in Fig. 11.5(b). The flow velocity is obtained by reading the voltmeter when the galvanometer (G) reading is zero after adjusting the variable

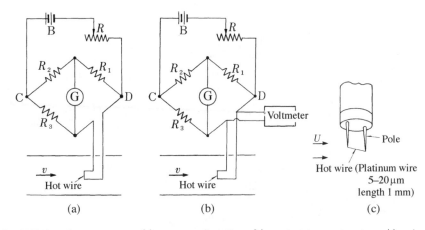

Fig. 11.5 Hot-wire anemometer: (a) constant voltage type; (b) constant temperature type; (c) probe

electrical resistance to maintain the hot-wire temperature, i.e. the electrical resistance, constant as the velocity changes. This is called the constant temperature anemometer (CTA).

Since the CTA has a good frequency response characteristic because thermal inertia effects are minimised, almost all currently used meters are of this type. It is capable of giving the flat characteristic up to a frequency of 100 kHz.

11.1.3 Laser Doppler anemometer

Point laser light at a tracer particle travelling with a fluid, and the scattered light from the particle develops a difference in frequency from the original incident light (reference light). This difference is due to the Doppler effect and is proportional to the particle velocity. A device by which the flow velocity is obtained from the velocity of tracer particles by measuring the difference in frequency f_D using a photocell or photodiode is called a laser Doppler anemometer.

Laser Doppler anemometers include the three types shown in Fig. 11.6 and described below.

Reference beam type
When a particle is moving in a fluid at velocity u as shown in Fig. 11.6(a), by measuring the difference in frequency f_D between the reference light and the scattered light observed in the direction of angle 2θ, the flow velocity u can be obtained from the following equation:

$$u = \frac{\lambda f_D}{2 \sin \theta} \tag{11.2}$$

where λ is the wavelength of the laser light.

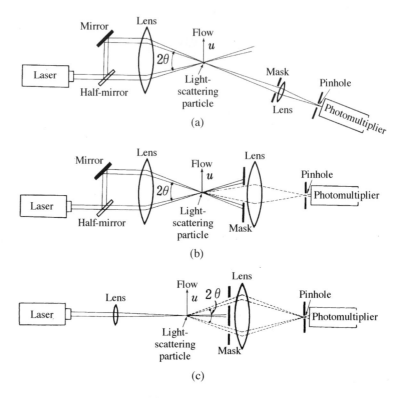

Fig. 11.6 Laser Doppler anemometers: (a) reference beam type; (b) interference fringe type; (c) single-beam type

Interference fringe type
As shown in Fig. 11.6(b), the flow velocity is obtained by using a photomultiplier to detect the alternating light intensity scattered when a particle passes the interference fringes. The velocity is again calculated using eqn (11.2).

Single-beam type
As shown in Fig. 11.6(c), by using the interference of the scattered light in two directions from a single incident beam, the flow velocity can be obtained as for the interference type.

11.2 Measurement of flow discharge

11.2.1 Method using a collecting vessel

This method involves measuring the fluid discharge by collecting it in a vessel and measuring its weight or volume. In the case of a gas, the temperature and pressure of the gas in the vessel are measured allowing conversion to another volume under standard conditions of temperature and pressure or to mass.

11.2.2 Methods using flow restrictions

Discharge measurement using flow restrictions is widely used in industry. Restrictions include the orifice, nozzle and Venturi tube. The flow rate is obtained by detecting the difference in pressures upstream and downstream of the device. Flow measurement methods are stipulated in British Standards BS1042 (1992).[1]

Orifice plate

The construction of an orifice plate is shown in Fig. 11.7. It is set inside a straight pipe. The flow rate is found by measuring the difference in pressures across it. The flow rate is calculated as follows:

$$Q = \alpha \frac{\pi d^2}{4} \sqrt{\frac{2\Delta p}{\rho}} \qquad (11.3)$$

where α is called the flow coefficient and Δp is the pressure difference across the orifice plate.

The symbol C was used for the coefficient of discharge in eqn (5.25). For

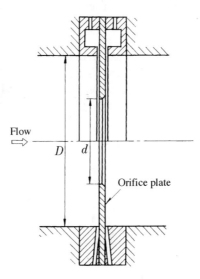

Flow

D d

Orifice plate

Fig. 11.7 Orifice plate with pressure tappings (corner and flange)

[1] British Standards BS1042, *Measurement of Fluid Flow in Closed Conduits*, British Standards Institution.

all the above cases, the relationship between flow coefficients α and coefficient of discharge C is

$$C = \alpha/E \qquad (11.4)$$

where the approach velocity coefficient $E = (1 - \beta^4)^{-1/2}$ and the throttle diameter ratio $\beta = d/D$.

It can be seen that the effect of the flow velocity in the pipe is to increase the flow rate for the same pressure drop Δp by the factor E, compared with flow from a tank or reservoir as in eqn (5.25).

To obtain the pressure difference either the corner tappings or flange tappings (Fig. 11.7) or pipe tappings are used.

For the case of a gas, an expansion factor is needed as follows:

$$Q_{v1} = \alpha\varepsilon\frac{\pi d^2}{4}\sqrt{\frac{2\Delta p}{\rho_1}} \qquad (11.5)$$

$$m = \alpha\varepsilon\frac{\pi d^2}{4}\sqrt{2\rho_1\Delta p} \qquad (11.6)$$

where Q_{v1} is the upstream volume flow rate, m is the mass flow rate, and ρ_1 is the upstream fluid density.

Nozzle

The design of a nozzle is shown in Fig. 11.8, and the measuring method and calculation formula are therefore the same as those for an orifice plate. For a nozzle, the flow loss is smaller than for an orifice, and also the flow coefficient is larger.

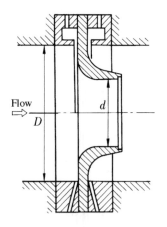

Fig. 11.8 ISA 1932 nozzle

Venturi tube

The principle of the Venturi tube was explained in Section 5.2.2. British Standards provides the standards for both nozzle-type and cone-type Venturi tubes as shown in Fig. 11.9.

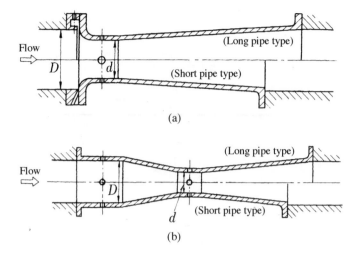

Fig. 11.9 Venturi tubes: (a) nozzle type; (b) cone type

The calculation of the discharge is again the same as that for the orifice plate:

$$Q = \alpha \frac{\pi d^2}{4} \sqrt{\frac{2\Delta p}{\rho}} \tag{11.7}$$

In the case of a gas, as for the orifice plate, eqns (11.4) and (11.5) are used.

11.2.3 Area flowmeter [2]

The flowmeters explained in Section 11.2.2 indicate the flow from the pressure difference across the restriction. An area flowmeter, however, has a changing level of restriction such that the pressure difference remains constant, and the flow rate is induced by the flow area. Area meters include float, piston and gate types.

A float-type area flowmeter (rotameter) has, as shown in Fig. 11.10, a float which is suspended in a vertical tapered tube. The flow produces a pressure difference across the float. The float rests in a position where the combined forces of pressure drag, frictional drag and buoyancy balance its weight. In this case, ignoring friction, flow Q is expressed by the following equation:

$$Q = C_d a_x \sqrt{\frac{2gV(\rho_f - \rho)}{\rho a_0}} \tag{11.8}$$

where ρ is the fluid density, C_d is the coefficient of discharge, a_x is the area

[2] British Standard BS7405, (1992).

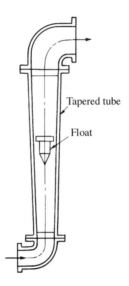

Fig. 11.10 Float-type area flowmeter (Rotameter)

of the annulus through which the fluid passes outside the float, V is the float volume, ρ_f is the float density and a_0 is the maximum section area of the float. Since a_x changes in proportion to the float position, if C_d is constant the equilibrium height of the float in the tube is proportional to the flow.

11.2.4 Positive displacement flowmeter

A positive displacement flowmeter with continuous flow relies on some form of measuring chamber of constant volume. It is then possible to obtain the integrated volume by counting the number of times the volume is filled, and the flow rate by measuring the number of times this is done per second. As a

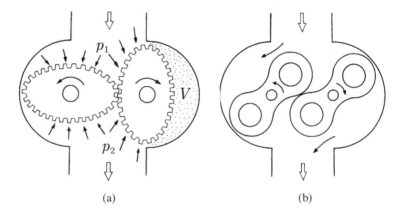

(a)   (b)

Fig. 11.11 Positive displacement flowmeters: (a) oval gear type; (b) Roots type

typical example, Fig. 11.11 shows oval gear and Roots-type positive displacement meters.

Because of the difference between the flow inlet pressure p_1 and the flow outlet pressure p_2 of fluid, the vertically set gear (Fig. 11.11(a)) turns in the direction of the arrow. Thus, every complete revolution sends out fluid of volume $4V$.

11.2.5 Turbine flowmeter

If a turbine is placed in the course of a flow, the turbine rotates owing to the velocity energy of the fluid. Since they are almost proportional, the flow velocity is obtainable from the rotational velocity of the turbine, while the integrated volume can be calculated by counting the number of revolutions.

The flowmeter has long been used as a water meter. Figure 11.12 shows a turbine meter used industrially for flow rate measurement of various fluids. A pulse is induced every time the blade of the turbine passes the magnetic coil face and the pulse frequency is proportional to the volume flow rate.

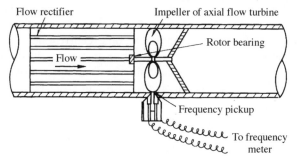

Fig. 11.12 Turbine flowmeter

11.2.6 Magnetic flowmeter

As shown in Fig. 11.13, when a conducting fluid flows in a non-conducting section of a measuring tube to which a magnetic field of flux density B is applied normal to the flow direction, an electromotive force E proportional to the mean flow velocity v is induced in the liquid (Faraday's law of electro-magnetic induction) which, after amplification, permits computation of the volume flow rate Q. The electromotive force is detected by inserting two electrodes into the tube in contact with the fluid and normal to both the flow and magnetic field directions. In other words, if the tube diameter is d, then

$$E = Bdv \qquad (11.9)$$

and

$$Q = \frac{\pi d E}{4B} \qquad (11.10)$$

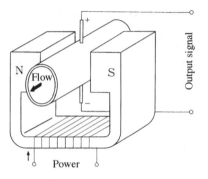

Fig. 11.13 Magnetic flowmeter

Since this flowmeter has no pressure loss, measurement can be made irrespective of the viscosity, specific gravity, pressure and Reynolds number of the fluid.

11.2.7 Ultrasonic flowmeter

As shown in Fig. 11.14, piezocrystals A and B are located a distance l apart on a line passing obliquely through the pipe centre line. Assume that an ultrasonic wave pulse sent from a transmitter at A is received by the detector at B t_1 seconds later. Then, exchanging the functions of A and B by the send–receive switch, an ultrasonic wave pulse sent from B is detected by A t_2 seconds later. Thus

$$t_1 = \frac{1}{a + v\cos\theta} \quad t_2 = \frac{1}{a - v\cos\theta}$$

$$\frac{1}{t_1} - \frac{1}{t_2} = \frac{a + v\cos\theta}{l} - \frac{a - v\cos\theta}{l} = \frac{2v\cos\theta}{l}$$

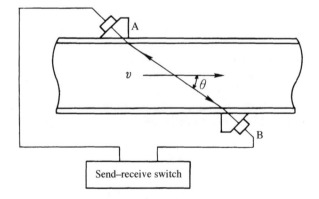

Fig. 11.14 Ultrasonic flowmeter

where a is the sonic velocity in the fluid. From this equation,

$$v = \frac{1}{2\cos\theta}\left(\frac{1}{t_1} - \frac{1}{t_2}\right) \tag{11.11}$$

This flowmeter has the same merits as an electromagnetic flowmeter and an additional benefit of usability in a non-conducting fluid. On the other hand it has the disadvantages of complex construction and high price.

11.2.8 Vortex shedding flowmeter [3]

If a cylinder (diameter d) is placed in a flow, Kármán vortices develop behind it. The frequency f of vortex shedding from the cylinder is shown in eqn (9.7). The Strouhal number changes with the Reynolds number, but it is almost constant at 0.2 within the range of $Re = 300$–$100\,000$. In other words, the flow velocity U is expressed by the following equation:

$$U = fd/0.2 \tag{11.12}$$

One practical configuration, shown in Fig. 11.15, induces fluid movement through the cylinder for electrical detection of the vortices, and thus measurement of the flow rate.

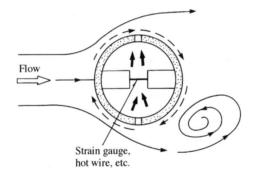

Fig. 11.15 Vortex shedding flowmeter

11.2.9 Fluidic flowmeter

As shown in Fig. 11.16, with an appropriate feedback mechanism a wall attachment amplifier can become a fluidic oscillator whose jet spontaneously oscillates at a frequency proportional to the volume flow rate of the main jet flow. The device can thus be used as a flowmeter.[4,5]

[3] Yamazaki, H. et al., Journal of Instrumentation and Control, 10 (1971), 173.
[4] Boucher, R. F. and Mazharoglu, C., International Gas Research Conference, (1987), 522.
[5] Yamazaki, H. et al., Proc. FLUCOME'85, Vol. 2 (1985), 617.

Fig. 11.16 Fluidic flowmeter

11.2.10 Weir [6]

As shown in Fig. 11.17, the three principal weir configurations are classified by shape into triangular, rectangular and full-width weirs.

Table 11.1 shows the flow computation formulae and the applicable scope for such weirs.

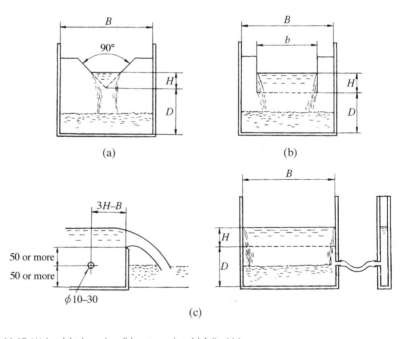

(a)

(b)

(c)

Fig. 11.17 Weirs: (a) triangular; (b) rectangular; (c) full width

[6] British Standard BS3680.

Table 11.1 Flow computation formulae for weirs in British Standard BS 3680

Kind of weir	Triangular weir	Rectangular weir		Full-width weir
Discharge computation formula	$Q = \frac{8}{15}C\sqrt{2g}H^{5/2}$ (m³/s) $C = 0.5785$ $H = H + 0.00085$	$Q = C\frac{2}{3}\sqrt{2g}bH^{3/2}$ (m³/s) $C = \left[0.578 + 0.037\left(\frac{b}{B}\right)^2 + \dfrac{0.003615 - 0.0030\left(\frac{b}{B}\right)}{H + 0.0016} \right]$ $\times\left[1 + 0.5\left(\frac{b}{B}\right)^4\left(\frac{H}{H+D}\right)^2 \right]$		$Q = C\frac{2}{3}\sqrt{2g}bH^{3/2}$ (m³/s) $C = 0.596 + 0.091\dfrac{H}{D}$ $H = H + 0.001$
Applicable range	$\dfrac{H}{D} < 0.4$ $\dfrac{H}{B} < 0.2$ $0.05\,\text{m} < H < 0.38\,\text{m}$ $D > 0.45\,\text{m}$ $B > 1.0\,\text{m}$	$\dfrac{b}{B} > 0.3$ $\dfrac{H}{D} < 1.0$ $0.025\,\text{m} < H < 0.80\,\text{m}$ $D > 0.30\,\text{m}$		$\dfrac{b}{B} = 1.0$ $\dfrac{H}{D} < 2.5$ $H > 0.03$ $b > 0.20$ $D > 0.10$

11.3 Problems

1. The velocity of water flowing in a pipe was measured with a Pitot tube, and the differential pressure read on a connected mercury manometer was 8 cm. Assuming that the velocity coefficient for the Pitot tube is one, obtain the flow velocity. Assume that the water temperature is 20°C and the specific gravity of mercury $s = 13.5$.

2. Air flow was measured with the three-hole Pitot tube shown in Fig. 11.18, and it was found that the heights B and C of the water manometer were equal whilst A was 5 cm lower. What was the air flow velocity? Assume that the temperature was 20°C and the air density is 1.205 kg/m³.

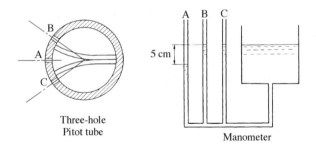

Three-hole
Pitot tube

Manometer

Fig. 11.18

3. An orifice of diameter 50 cm on a pipe of diameter 100 mm was used to measure air flow. The differential pressure read on a connected mercury manometer was 120 mm. Assuming that the discharge coefficient $\alpha = 0.62$ and the gas expansion coefficient $\varepsilon = 0.98$, obtain the mass flow

rate. Assume that the pressure and temperature upstream of the orifice are 196 kPa and 20°C respectively.

4. A volume of $3.6 \times 10^{-3}\,\mathrm{m}^3$ of water per second flows from an orifice of diameter $d = 50\,\mathrm{mm}$ in the side of the water tank shown in Fig. 11.19. The minimum section diameter of the jet flow is $d' = 40\,\mathrm{mm}$. Obtain the contraction coefficient C_c, velocity coefficient C_v and discharge coefficient C of this orifice.

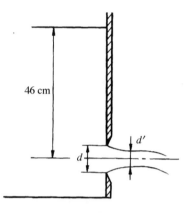

46 cm

d'

d

Fig. 11.19

5. A pipe line contains both an orifice and a nozzle. When $Re = 1 \times 10^5$, and with a throttle diameter ratio $\beta = 0.6$ for both, the flow coefficient α is 0.65 for the orifice but 1.03 for the nozzle. Explain why.

6. Explain the principle of a hot-wire flow anemometer. Over what points should caution be especially exercised?

7. Explain the principles and features of the laser Doppler anemometer.

8. With a vortex-shedding flowmeter, cylinder diameter 2 cm, the shedding frequency was measured as 5 Hz. What was the flow velocity?

9. Obtain the flow formulae for a rectangular weir and a triangular weir.

10. Assuming a reading error of 2% for both rectangular and triangular weirs, what are the resulting percentage flow errors?

12

Flow of an ideal fluid

When the Reynolds number Re is large, since the diffusion of vorticity is now small (eqn (6.18)) because the boundary layer is very thin, the overwhelming majority of the flow is the main flow. Consequently, although the fluid itself is viscous, it can be treated as an ideal fluid subject to Euler's equation of motion, so disregarding the viscous term. In other words, the applicability of ideal flow is large.

For an irrotational flow, the velocity potential ϕ can be defined so this flow is called the potential flow. Originally the definition of potential flow did not distinguish between viscous and non-viscous flows. However, now, as studied below, potential flow refers to an ideal fluid.

In the case of two-dimensional flow, a stream function ψ can be defined from the continuity equation, establishing a relationship where the Cauchy–Riemann equation is satisfied by both ϕ and ψ. This fact allows theoretical analysis through application of the theory of complex variables so that ϕ and ψ can be obtained. Once ϕ or ψ is obtained, velocities u and v in the x and y directions respectively can be obtained, and the nature of the flow is revealed.

In the case of three-dimensional flow, the theory of complex variables cannot be used. Rather, Laplace's equation $\Delta^2\phi = 0$ for a velocity potential $\phi = 0$ is solved. Using this approach the flow around a sphere etc. can be determined.

Here, however, only two-dimensional flows will be considered.

12.1 Euler's equation of motion

Consider the force acting on the small element of fluid in Fig. 12.1. Since the fluid is an ideal fluid, no force due to viscosity acts. Therefore, by Newton's second law of motion, the sum of all forces acting on the element in any direction must balance the inertia force in the same direction. The pressure acting on the small element of fluid $dx\,dy$ is, as shown in Fig. 12.1, similar to Fig. 6.3(b). In addition, taking account of the body force and also assuming that the sum of these two forces is equal to the inertial force, the equation of motion for this case can be obtained. This is the case where the

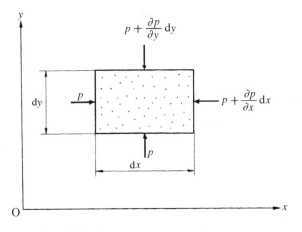

Fig. 12.1 Balance of pressures on fluid element

viscous term of eqn (6.12) is omitted. Consequently the following equations are obtainable:

$$\left.\begin{array}{l}\rho\left(\dfrac{\partial u}{\partial t}+u\dfrac{\partial u}{\partial x}+v\dfrac{\partial u}{\partial y}\right)=\rho X-\dfrac{\partial p}{\partial x}\\[2mm]\rho\left(\dfrac{\partial v}{\partial t}+u\dfrac{\partial v}{\partial x}+v\dfrac{\partial v}{\partial y}\right)=\rho Y-\dfrac{\partial p}{\partial y}\end{array}\right\}\qquad(12.1)$$

These are similar equations to eqn (5.4), and are called Euler's equations of motion for two-dimensional flow.

For a steady flow, if the body force term is neglected, then:

$$\left.\begin{array}{l}\rho\left(u\dfrac{\partial u}{\partial x}+v\dfrac{\partial u}{\partial y}\right)=-\dfrac{\partial p}{\partial x}\\[2mm]\rho\left(u\dfrac{\partial v}{\partial x}+v\dfrac{\partial v}{\partial y}\right)=-\dfrac{\partial p}{\partial y}\end{array}\right\}\qquad(12.2)$$

If u and v are known, the pressure is obtainable from eqn (12.1) or eqn (12.2).

Generally speaking, in order to obtain the flow of an ideal fluid, the continuity equation (6.2) and the equation of motion (12.1) or eqn (12.2) must be solved under the given initial conditions and boundary conditions. In the flow fluid, three quantities are to be obtained, namely u, v and p, as functions of t and x, y. However, since the acceleration term, i.e. inertial term, is non-linear, it is so difficult to obtain them analytically that a solution can only be obtained for a particular restricted case.

12.2 Velocity potential

The velocity potential ϕ as a function of x and y will be studied. Assume that

$$u = \frac{\partial \phi}{\partial x} \qquad v = \frac{\partial \phi}{\partial y} \qquad (12.3)^1$$

From $\partial u/\partial y = \partial^2 \phi/\partial y \partial x = \partial^2 \phi/\partial x \partial y = \partial v/\partial x$ the following relationship is obtained:

$$\frac{\partial u}{\partial y} - \frac{\partial v}{\partial x} = 0 \qquad (12.4)$$

This is the condition for irrotational motion. Conversely, if a flow is irrotational, function ϕ as in the following equation must exist for u and v:

$$d\phi = u\,dx + v\,dy \qquad (12.5)$$

Using eqn (12.3),

$$d\phi = \frac{\partial \phi}{\partial x}dx + \frac{\partial \phi}{\partial y}dy \qquad (12.6)$$

Consequently, when the function ϕ has been obtained, velocities u and v can also be obtained by differentiation, and thus the flow pattern is found. This function ϕ is called velocity potential, and such a flow is called potential or irrotational flow. In other words, the velocity potential is a function whose gradient is equal to the velocity vector.

Equation (12.6) turns out as follows if expressed in polar coordinates:

$$v_r = \frac{\partial \phi}{\partial r} \qquad v_\theta = \frac{\partial \phi}{r\,\partial \theta} \qquad (12.7)$$

For the potential flow of an incompressible fluid, substitute eqn (12.3) into continuity equations (6.2), and the following relationship is obtained:

$$\frac{\partial^2 \phi}{\partial x^2} + \frac{\partial^2 \phi}{\partial y^2} = 0 \qquad (12.8)^2$$

Equation (12.8), called Laplace's equation, is thus satisfied by the velocity potential used in this manner to express the continuity equation. From any solution which satisfies Laplace's equation and the particular boundary conditions, the velocity distribution can be determined. It is particularly

---

[1] In general, whenever u, v and w are respectively expressed as $\partial \phi/\partial x$, $\partial \phi/\partial y$ and $\partial \phi/\partial z$ for vector $V(x, y$ and z components are respectively u, v and $w)$, vector V is written as grad ϕ or $\nabla \phi$:

$$V = \text{grad}\,\phi = \nabla \phi = \left[\frac{\partial \phi}{\partial x}, \frac{\partial \phi}{\partial y}, \frac{\partial \phi}{\partial z}\right]$$

Equation (12.3) is the case of two-dimensional flow where $w = 0$, and can be written as grad ϕ or $\nabla \phi$.

[2] That is

$$\text{div}\,V = \text{div}[u, v, w] = \text{div}(\text{grad}\,\phi) = \text{div}\,\nabla \phi = \text{div}\left[\frac{\partial \phi}{\partial x}, \frac{\partial \phi}{\partial y}, \frac{\partial \phi}{\partial z}\right] = \frac{\partial u}{\partial x} + \frac{\partial v}{\partial y} + \frac{\partial w}{\partial z}$$

$$= \frac{\partial^2 \phi}{\partial x^2} + \frac{\partial^2 \phi}{\partial y^2} + \frac{\partial^2 \phi}{\partial z^2}$$

$\partial^2/\partial x^2 + \partial^2/\partial y^2 + \partial^2/\partial z^2$ is called the Laplace operator (Laplacian), abbreviated to Δ. Equation (12.8) is for a two-dimensional flow where $w = 0$, expressed as $\Delta \phi = 0$.

noteworthy that the pattern of potential flow is determined solely by the continuity equation and the momentum equation serves only to determine the pressure.

A line along which ϕ has a constant value is called the equipotential line, and on this line, since $d\phi = 0$ and the inner product of both vectors of velocity and the tangential line is zero, the direction of fluid velocity is at right angles to the equipotential line.

12.3 Stream function

For incompressible flow, from the continuity equation (6.2),

$$\frac{\partial u}{\partial x} + \frac{\partial v}{\partial y} = 0 \qquad (12.9)$$

This is eqn (12.4) but with u and v respectively replaced by $-v$ and u. Consequently, corresponding to eqn (12.5), it turns out that there exists a function ψ for x and y shown by the following equation:

$$d\psi = -v\,dx + u\,dy \qquad (12.10)$$

In general, since

$$d\psi = \frac{\partial \psi}{\partial x}dx + \frac{\partial \psi}{\partial y}dy \qquad (12.11)$$

u and v are respectively expressed as follows:

$$-v = \frac{\partial \psi}{\partial x} \qquad u = \frac{\partial \psi}{\partial y} \qquad (12.12)$$

Consequently, once function ψ has been obtained, differentiating it by x and y gives velocities v and u, revealing the detail of the fluid motion. ψ is called the stream function.

Expressing the above equation in polar coordinates gives

$$v_r = \frac{\partial \psi}{r\,\partial \theta} \qquad v_\theta = -\frac{\partial \psi}{\partial r} \qquad (12.13)$$

In general, for two-dimensional flow, the streamline is as follows, from eqn (4.1):

$$\frac{dx}{u} = \frac{dy}{v}$$

or

$$-v\,dx + u\,dy = 0 \qquad (12.14)$$

From eqns (12.12) and (12.14), the corresponding $d\psi = 0$, i.e. $\psi = $ constant, defines a streamline. The product of the tangents of a streamline and an equipotential line at the crossing point of both lines is as follow from eqns (12.3) and (12.12):

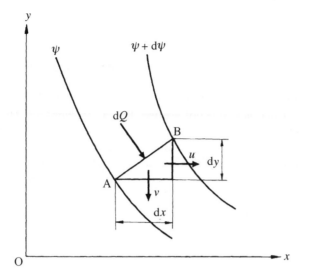

Fig. 12.2 Relationship between flow rate and stream function

$$\left(\frac{dy}{dx}\right)_\phi \left(\frac{dy}{dx}\right)_\psi = \left(\frac{\partial \psi}{\partial x}\bigg/\frac{\partial \psi}{\partial y}\right) \times \left(\frac{\partial \phi}{\partial x}\bigg/\frac{\partial \phi}{\partial y}\right) = -1$$

This relationship shows that the streamline intersects normal to the equipotential line at the crossing point of the two lines.

As shown in Fig. 12.2, consider points A and B on two closely neighbouring streamlines, ψ and $\psi + d\psi$. The volume flow rate dQ flowing in unit time and crossing line AB is as follows from the figure:

$$dQ = u\,dy - v\,dx = \frac{\partial \psi}{\partial y}dy + \frac{\partial \psi}{\partial x}dx = d\psi$$

The volume flow rate Q of fluid flowing between two streamlines $\psi = \psi_1$ and $\psi = \psi_2$ is thus given by the following equation:

$$Q = \int dQ = \int_{\phi_1}^{\phi_2} d\psi = \psi_2 - \psi_1 \tag{12.15}$$

Substituting eqn (12.12) into (4.8) for flow without vorticity, the following is obtained, clarifying that the stream function satisfies Laplace's equation:

$$\frac{\partial^2 \psi}{\partial x^2} + \frac{\partial^2 \psi}{\partial y^2} = 0 \tag{12.16}$$

12.4 Complex potential

For a two-dimensional incompressible potential flow, since the velocity potential ϕ and stream function ψ exist, the following equations result from eqns (12.3) and (12.12):

$$\frac{\partial \phi}{\partial x} = \frac{\partial \psi}{\partial y} \qquad \frac{\partial \phi}{\partial y} = -\frac{\partial \psi}{\partial x} \qquad (12.17)$$

These equations are called the Cauchy–Riemann equations in the theory of complex variables. In this case they express the relationship between the velocity potential and stream function. The Cauchy–Riemann equations clarify the fact that ϕ and ψ both satisfy Laplace's equation. They also clarify the fact that a combination of ϕ and ψ satisfying the Cauchy–Riemann conditions expresses a two-dimensional incompressible potential flow.

Now, consider a regular function[3] $w(z)$ of complex variable $z = x + iy$ and express it as follows by dividing it into real and imaginary parts:

$$\left. \begin{array}{l} w(z) = \phi + i\psi \\ z = x + iy = r(\cos\theta + i\sin\theta) = re^{i\theta} \\ \phi = \phi(x, y) \qquad \psi = \psi(x, y) \end{array} \right\} \qquad (12.18)$$

and ϕ and ψ above satisfy eqn (12.17) owing to the nature of a regular function. Consequently, real part $\phi(x, y)$ and imaginary part $\psi(x, y)$ of the regular function $w(z)$ of complex number z can respectively be regarded as the velocity potential and the stream function of the two-dimensional incompressible potential flow. In other words, there exists an irrotational motion whose equipotential line is $\phi(x, y) = $ constant and streamline $\psi(x, y) = $ constant. Such a regular function $w(z)$ is called the complex potential.

From eqn (12.18)

$$dw = \frac{\partial w}{\partial x}dx + \frac{\partial w}{\partial y}dy = \left(\frac{\partial \phi}{\partial x} + i\frac{\partial \psi}{\partial x}\right)dx + \left(\frac{\partial \phi}{\partial y} + i\frac{\partial \psi}{\partial y}\right)dy$$
$$= (u - iv)dx + (v + iu)dy = (u - iv)(dx + i\,dy) = (u - iv)dz$$

Therefore

$$\frac{dw}{dz} = u - iv \qquad (12.19)$$

Consequently, whenever $w(z)$ is differentiated with respect to z, as shown in Fig. 12.3, its real part yields velocity u in the x direction, and the negative of its imaginary part yields velocity v in the y direction. The actual velocity $u + iv$ is called the complex velocity while $u - iv$ in the above equation is the conjugate complex velocity.

---

[3] The function whose differential at any point with respect to z is independent of direction in the z plane is called a regular function. A regular function satisfies the Cauchy–Riemann equations.

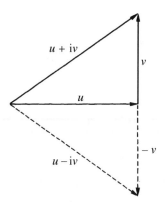

Fig. 12.3 Complex velocity

12.5 Example of potential flow

12.5.1 Basic example

Parallel flow

For the uniform flow U shown in Fig. 12.4, from eqn (12.3)

$$u = \frac{\partial \phi}{\partial x} = U \quad v = \frac{\partial \phi}{\partial y} = 0$$

Therefore

$$d\phi = \frac{\partial \phi}{\partial x} dx + \frac{\partial \phi}{\partial y} dy = U\, dx$$
$$\phi = Ux$$

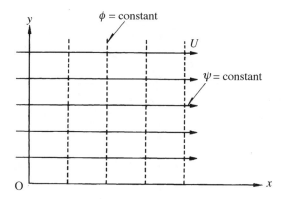

Fig. 12.4 Parallel flow

From eqn (12.12)

$$u = \frac{\partial \psi}{\partial y} = U \quad v = -\frac{\partial \psi}{\partial x} = 0$$

Therefore

$$d\phi = \frac{\partial \psi}{\partial x} dx + \frac{\partial \psi}{\partial y} dy = U \, dy$$

$$\psi = Uy$$

$$w(z) = \phi + i\psi = U(x + iy) = Uz \qquad (12.20)$$

The complex potential of parallel flow U in the x direction emerges as $w(z) = uz$.

Furthermore, if the complex potential is given as $w(z) = Uz$, the conjugate complex velocity is

$$\frac{dw}{dz} = U \qquad (12.21)$$

clarifying again that it expresses a uniform flow in the direction of the x axis.

Source
As shown in Fig. 12.5, consider a case where fluid discharges from the origin (point O) at quantity q per unit time. Putting velocity in the radial direction on a circle of radius r to v_r, the discharge q per unit thickness is

$$q = 2\pi r v_r = \text{constant} \qquad (12.22)$$

From eqns (12.7) and (12.22)

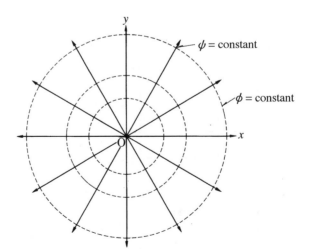

Fig. 12.5 Source

$$v_r = \frac{\partial \phi}{\partial r} = \frac{q}{2\pi r}$$

Also, from eqn (12.7),

$$v_\theta = \frac{\partial \phi}{r\, \partial \theta} = 0$$

Integrating $d\phi$ in the above equation gives

$$\phi = \frac{q}{2\pi} \log r \qquad (12.23)$$

Then, from eqns (12.13) and (12.22),

$$v_r = \frac{\partial \psi}{r\, \partial \theta} = \frac{q}{2\pi r} \qquad v_\theta = -\frac{\partial \psi}{\partial r} = 0$$

Therefore

$$\psi = \frac{q}{2\pi}\theta \qquad (12.24)$$

Consequently, the complex potential is expressed by the following equation:

$$w = \phi + i\psi = \frac{q}{2\pi}(\log r + i\theta) = \frac{q}{2\pi}\log(re^{i\theta}) = \frac{q}{2\pi}\log z \qquad (12.25)$$

From eqns (12.23) and (12.24) it is known that the equipotential lines are a set of circles centred at the origin while the streamlines are a set of radial lines radiating from the origin. Also, it is noted that the flow velocity v_r is inversely proportional to the distance r from the origin.

Whenever $q > 0$, fluid flows out evenly from the origin towards the periphery. Such a point is called a source. Conversely, whenever $q < 0$, fluid is absorbed evenly from the periphery. Such a point is called a sink. $|q|$ is called the strength of the source or sink.

Free vortex
In Fig. 12.6, fluid rotates around the origin with tangential velocity v_θ at any given radius r. The circulation Γ is as follows from eqn (4.9):

$$\tilde{A} = \int_{\theta=0}^{2\pi} v_\theta\, ds = v_\theta r \int_0^{2\pi} d\theta = 2\pi r v_\theta$$

The velocity potential ϕ is

$$v_\theta = \frac{\partial \phi}{r\, \partial \theta} = \frac{\tilde{A}}{2\pi r} \qquad v_r = \frac{\partial \phi}{\partial r} = 0$$

Therefore

$$\phi = \frac{\Gamma}{2\pi}\theta \qquad (12.26)$$

It emerges that v_θ is inversely proportional to the distance from the centre.
The stream function ψ is

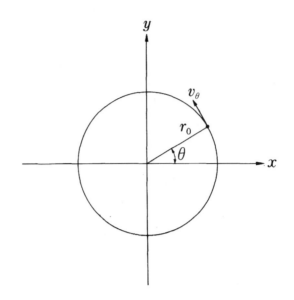

Fig. 12.6 Vortex

$$v_\theta = -\frac{\partial \psi}{\partial r} = \frac{\Gamma}{2\pi r} \quad v_r = \frac{\partial \psi}{r\,\partial \theta} = 0$$

Therefore

$$\psi = -\frac{\Gamma}{2\pi}\log r \tag{12.27}$$

Consequently, the complex potential is

$$w(z) = \phi + \mathrm{i}\psi = \frac{\Gamma}{2\pi}(\theta = \mathrm{i}\log r) = -\frac{\mathrm{i}\Gamma}{2\pi}(\log r + \mathrm{i}\theta) = -\frac{\mathrm{i}\Gamma}{2\pi}\log z \tag{12.28}$$

For clockwise circulation, $w(z) = (\mathrm{i}\Gamma/2\pi)$.

From eqns (12.26) and (12.17), it is known that the equipotential lines are a group of radial straight lines passing through the origin whilst the flow lines are a group of concentric circles centred on the origin. This flow appears in Fig. 12.5 with broken lines representing streamlines and solid lines as equipotential lines. The circulation Γ is positive counterclockwise, and negative clockwise.

This flow consists of rotary motion in concentric circles around the origin with the velocity inversely proportional to the distance from the origin. Such a flow is called a free vortex while the origin point itself is a point vortex. The circulation is also called the strength of the vortex.

12.5.2 Synthesising of flows

When there are two regular functions $w_1(z)$ and $w_2(z)$, the function obtained as their sum

$$w(z) = w_1(z) + w_2(z) \tag{12.29}$$

is also a regular function. If w_1 and w_2 represent the complex potentials of two flows, another complex potential is obtained from their sum. By combining two two-dimensional incompressible potential flows in such a manner, another flow can be obtained.

Combining a source and a sink

Assume that, as shown in Fig. 12.7, the source q is at point A ($z = -a$) and sink $-q$ is at point B ($z = a$).

The complex potential w_1 at any point z due to the source whose strength is q at point A is

$$w_1 = \frac{q}{2\pi} \log(z + a) \qquad (12.30)$$

The complex potential w_2 at any point z due to the sink whose strength is q is

$$w_2 = -\frac{q}{2\pi} \log(z - a) \qquad (12.31)$$

Because of the linearity of Laplace's equation the complex potential w of the flow which is the combination of these two flows is

$$w = \frac{q}{2\pi} [\log(z + a) - \log(z - a)] \qquad (12.32)$$

Now, from Fig. 12.7, since

$$z + a = r_1 e^{i\theta_1} \qquad z - a = r_2 e^{i\theta_2}$$

from eqn (12.32)

$$w = \frac{q}{2\pi} \left(\log \frac{r_1}{r_2} + i(\theta_1 - \theta_2) \right) \qquad (12.33)$$

Therefore

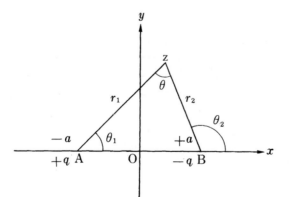

Fig. 12.7 Definition of variables for source A and sink B combination

$$\phi = \frac{q}{2\pi} \log\left(\frac{r_1}{r_2}\right) \tag{12.34}$$

$$\psi = \frac{q}{2\pi}(\theta_1 - \theta_2) \tag{12.35}$$

Assuming $\phi = $ constant from the first equation, equipotential lines are obtainable which are Appolonius circles for points A and B (a group of circles whose ratios of distances from fixed points A and B are constant). Taking $\psi = $ constant, streamlines are obtainable which are found to be another set of circles whose vertical angles are the constant angle $(\theta_1 - \theta_2)$ for chord AB (Fig. 12.8).

Consider the case where $a \to 0$ in Fig. 12.8, under the condition of $aq = $ constant. Then from eqn (12.32),

$$w = \frac{q}{2\pi} \log\left(\frac{1 + a/z}{1 - a/z}\right) = \frac{q}{\pi}\left[\frac{a}{z} + \frac{1}{3}\left(\frac{a}{z}\right)^3 + \frac{1}{5}\left(\frac{a}{z}\right)^5 + \cdots\right] = \frac{aq}{\pi z} = \frac{m}{z} \tag{12.36}$$

A flow given by the complex potential of eqn (12.36) is called a doublet, while $m = aq/\pi$ is its strength. The concept of a doublet is the extremity of a source and a sink of equal strength approaching infinitesimally close to each other whilst increasing their strength.

From eqn (12.36),

$$w = \frac{m}{x + iy} = m\frac{x - iy}{x^2 + y^2} \tag{12.37}$$

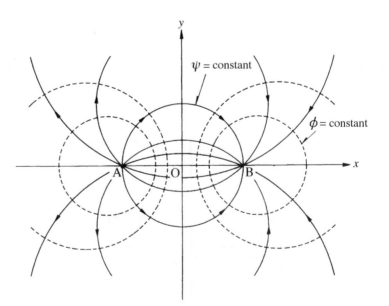

Fig. 12.8 Flow due to the combination of source and sink

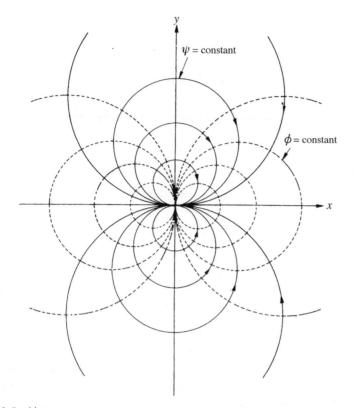

$\psi = $ constant

$\phi = $ constant

Fig. 12.9 Doublet

$$\phi = \frac{mx}{x^2 + y^2} \qquad (12.38)$$

$$\psi = \frac{my}{x^2 + y^2} \qquad (12.39)$$

From these equations, as shown in Fig. 12.9, an equipotential line is a circle whose centre is on the x axis whilst being tangential to the y axis, and a streamline is a circle whose centre is on the y axis whilst being tangential to the x axis.

Flow around a cylinder
Consider a circle of radius r_0 centred at the origin in uniform parallel flows. In general, by placing a number of sources and sinks in parallel flows, flows around variously shaped bodies are obtainable. In this case, however, by superimposing parallel flows onto the same doublet shown in Fig. 12.9, flows around a circle are obtainable as follows.

From eqns (12.29) and (12.36) the complex potential when a doublet is in uniform flows U is

$$w(z) = Uz + \frac{m}{z} = U\left(z + \frac{m}{U}\frac{1}{z}\right)$$

Now, put $m/U = r_0^2$, and

$$w(z) = U\left(z + \frac{r_0^2}{z}\right) \tag{12.40}$$

Decompose the above using the relationship $z = r(\cos\theta + i\sin\theta)$, and

$$w(z) = U\left(r + \frac{r_0^2}{r}\right)\cos\theta + iU\left(r - \frac{r_0^2}{r}\right)\sin\theta$$

$$\phi = U\left(r + \frac{r_0^2}{r}\right)\cos\theta \tag{12.41}$$

$$\psi = U\left(r - \frac{r_0^2}{r}\right)\sin\theta \tag{12.42}$$

Also, the conjugate complex velocity is

$$\frac{dw}{dz} = U - \frac{Ur_0^2}{z^2} \tag{12.43}$$

with stagnation points at $z = \pm r_0$. The streamline passing the stagnation point $\psi = 0$ is given by the following equation:

$$\left(r - \frac{r_0^2}{r}\right)\sin\theta = 0$$

This streamline consists of the real axis and the circle of radius r_0 centred at the origin. By replacing this streamline with a solid surface, the flow around a cylinder is obtained as shown in Fig. 12.10.

The tangential velocity of flow around a cylinder is, from eqn (12.41),

$$v_\theta = \frac{1}{r}\frac{\partial\phi}{\partial\theta} = -U\left(1 + \frac{r_0^2}{r^2}\right)\sin\theta \tag{12.44}$$

Since $r = r_0$ on the cylinder surface,

$$v_\theta = -2U\sin\theta$$

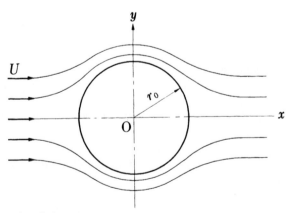

Fig. 12.10 Flow around a cylinder

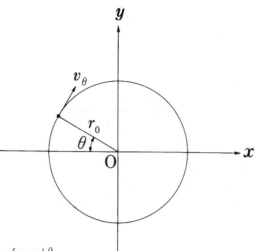

Fig. 12.11 Definitions of v_θ and θ

When the directions of θ and v_θ are arranged as shown in Fig. 12.11, this becomes

$$v_\theta = 2U \sin \theta \qquad (12.45)$$

The complex potential when there is clockwise circulation Γ around the cylinder is, as follows from eqns (12.28) and (12.40),

$$w(z) = U\left(z + \frac{r_0^2}{z}\right) + \frac{i\Gamma}{2\pi}\log z \qquad (12.46)$$

The flow in this case turns out as shown in Fig. 12.12. The tangential velocity v_θ' on the cylinder surface is as follows:

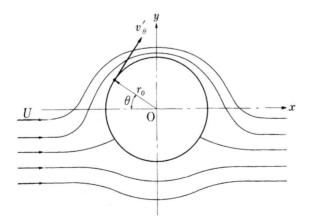

Fig. 12.12 Flow around a cylinder with circulation

$$v'_\theta = 2U \sin \theta + \frac{\Gamma}{2\pi r_0} \tag{12.47}$$

12.6 Conformal mapping

A simple flow can be studied within the limitations of the z plane as in the preceding section. For a complex flow, however, there may be some established cases of useful mapping of a transformation to another plane. For example, by transforming flow around a cylinder etc. through mapping functions onto some other planes, such complex flows as the flow around a wing, and between the blades of a pump, blower or turbine, can be determined.

Assume that there is the relationship

$$\zeta = f(z) \tag{12.48}$$

between two complex variables $z = x + iy$ and $\zeta = \xi + i\eta$, and that ζ is the regular function of z. Consider a mesh composed of $x = $ constant and $y = $ constant on the z plane as shown in Fig. 12.13. That mesh transforms to another mesh composed of $\xi = $ constant and $\eta = $ constant on the ζ plane. In other words, the pattern on the z plane is different from the pattern on the ζ plane but they are related to each other.

Further, assume that, as shown in Fig. 12.14, point ζ_0 corresponds to point z_0 and that the points corresponding to points z_1 and z_2 both minutely off z_0 are ζ_1 and ζ_2. Then

$$z_1 - z_0 = r_1 e^{i\theta_1} \qquad z_2 = z_0 = r_2 e^{i\theta_2}$$
$$\zeta_1 - \zeta_2 = R_1 e^{i\beta_1} \qquad \zeta_2 - \zeta_0 = R e^{i\beta_2}$$

From eqn (12.48),

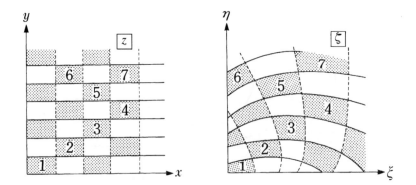

Fig. 12.13 Corresponding mesh on ζ and z planes

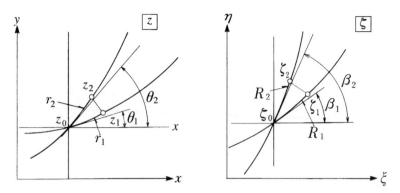

Fig. 12.14 Conformal mapping

$$\lim_{z_1 \to z_2} \left(\frac{\zeta_1 - \zeta_0}{z_1 - z_0} \right) = \left(\frac{d\zeta}{dz} \right)_{z=z_0} = \lim_{z_1 \to z_2} \left(\frac{\zeta_2 - \zeta_0}{z_2 - z_0} \right)$$

or

$$\frac{R_1 e^{i\beta_1}}{r_1 e^{i\theta_1}} = \frac{R_2 e^{i\beta_2}}{r_2 e^{i\theta_2}}$$

From the above, it turns out that

$$\frac{r_2}{r_1} = \frac{R_2}{R_1} \quad \theta_2 - \theta_1 = \beta_2 - \beta_1$$

and the minute triangles on the z plane are

$$\Delta z_0 z_1 z_2 \propto \Delta \zeta_0 \zeta_1 \zeta_2 \tag{12.49}$$

This shows that even though the pattern as a whole on the z plane may be very different from that on the ζ plane, their minute sections are similar and equiangularly mapped. Such a manner of pattern mapping is called conformal mapping, and $f(z)$ is the mapping function.

Now, consider the mapping function

$$\zeta = z + \frac{a^2}{z} \quad (a > 0) \tag{12.50}$$

Substitute a circle of radius a on the z plane, $z = ae^{i\theta}$, into eqn (12.50),

$$\zeta = a(e^{i\theta} + 1/e^{i\theta}) = a(e^{i\theta} + e^{-i\theta}) = 2a \cos\theta \tag{12.51}$$

At the time when θ changes from 0 to 2π, ζ corresponds in $2a \to 0 \to -2a \to 0 \to 2a$. In other words, as shown in Fig. 12.15(a), the cylinder on the z plane is conformally mapped onto the flat board on the ζ plane. The mapping function in eqn (12.50) is renowned, and is called Joukowski's transformation.

If conformal mapping is made onto the ζ plane using Joukowski's mapping function (12.50) while changing the position and size of a cylinder on the z plane, the shape on the ζ plane changes variously as shown in Fig. 12.15.

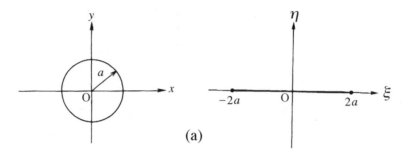

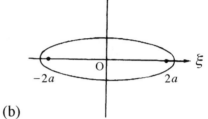

(a)

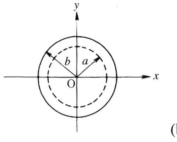

(b)

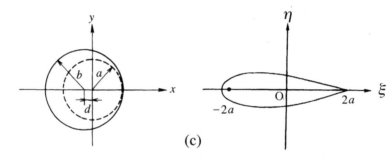

(c)

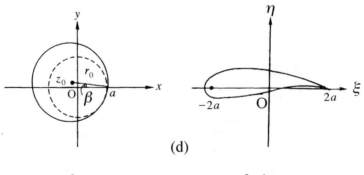

(d)

z plane   ζ plane

Fig. 12.15 Mapping of cylinders through Joukowski's transformation: (a) flat plate; (b) elliptical section; (c) symmetrical wing; (d) asymmetrical wing

The flow around the asymmetrical wing appearing in Fig. 12.15(d) can be obtained by utilising Joukowski's conversion. Consider the flow in the case where a cylinder of eccentricity z_0 and radius r_0 is placed in a uniform flow U whose circulation strength is Γ. The complex potential of this flow can be obtained by substituting $z - z_0$ for z in eqn (12.46),

$$w = U\left((z - z_0) + \frac{r_0^2}{z - z_0}\right) + i\frac{\Gamma}{2\pi}\log(z - z_0) \qquad (12.52)$$

Putting $z = z_0 + re^{i\theta}$, from $w = \phi + i\psi$

$$\phi = U\left(r + \frac{r_0^2}{r}\right)\cos\theta - \frac{\Gamma}{2\pi}\theta \qquad (12.53)$$

$$\psi = U\left(r - \frac{r_0^2}{r}\right)\sin\theta - \frac{\Gamma}{2\pi}\log r \qquad (12.54)$$

On the circle $r = r_0$, $\psi = $ constant, comprising a streamline. According to the Kutta condition[4] (where the trailing edge must become a stagnation point),

$$\left(\frac{d\phi}{d\theta}\right)_{\substack{\theta = -\beta \\ r = r_0}} = 2Ur_0 \sin\beta - \frac{\Gamma}{2\pi} = 0 \qquad (12.55)$$

Therefore

$$\Gamma = 4\pi Ur_0 \sin\beta \qquad (12.56)$$

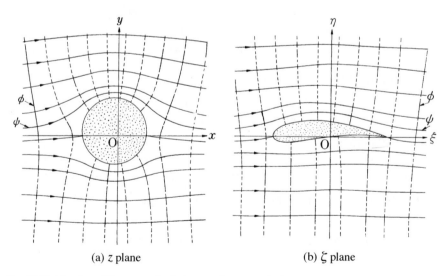

(a) z plane   (b) ζ plane

Fig. 12.16 Mapping of flow around cylinder onto flow around wing

---

[4] If the trailing edge was not a stagnation point, the flow would go around the sharp edge at infinite velocity from the lower face of the wing towards the upper face. The Kutta condition avoids this physical impossibility.

Equipotential lines and streamlines produced by substituting values of Γ satisfying eqn (12.56) into eqns (12.53) and (12.54) are shown in Fig. 12.16(a). They can be conformally mapped onto the ζ plane by utilising Joukowski's conversion by eliminating z from eqns (12.50) and (12.52) to obtain the complex potential on the ζ plane. The resulting flow pattern around a wing can be found as shown in Fig. 12.16(b). In this way, by means of conformal mapping of simple flows, such as around a cylinder, flow around complex-shaped bodies can be found.

Since the existence of analytical functions which shift z to the outside territory of given wing shapes is generally known, the behaviour of flow around these wings can be found from the flow around a cylinder through a process similar to the previous one. In addition, there are examples where it can be used for computing the contraction coefficient[5] of flow out of an orifice in a large vessel and the drag[6] due to the flow behind a flat plate normal to the flow.

12.7 Problems

1. Obtain the velocity potential and the flow function for a flow whose components of velocity in the x and y directions at a given point in the flow are u_0 and v_0 respectively.

2. Show the existence of the following relationship between flow function ψ and the velocity components v_r, v_θ in a two-dimensional flow:

$$v_\theta = -\frac{\partial \psi}{\partial r} \qquad v_r = \frac{\partial \psi}{r\, \partial \theta}$$

3. What is the flow whose velocity potential is expressed as $\phi = \Gamma\theta/2\pi$?

4. Obtain the velocity potential and the stream function for radial flow from the origin at quantity q per unit time.

5. Assuming that $\psi = U(r - r_0^2/r)\sin\theta$ expresses the stream function around a cylinder of radius r_0 in a uniform flow of velocity U, obtain the velocity distribution and the pressure distribution on the cylinder surface.

6. Obtain the pattern of flow whose complex potential is expressed as $w = x^2$.

7. What is the flow expressed by the following complex potential?

$$w(z) = \phi + i\psi = \frac{i\Gamma}{2\pi}\log z$$

---

[5] Lamb, H., *Hydrodynamics*, (1932), 6th edition, 98, Cambridge University Press.
[6] Kirchhoff, G., *Grelles Journal*, 70 (1869), 289.

8. Obtain the complex potential of a uniform flow at angle α to the x axis.

9. Obtain the streamline $y = k$ and the equipotential line $x = c$ of a flow parallel to the x axis on the z plane when mapped onto the ζ plane by mapping function $\zeta = 1/z$.

10. Obtain the flow in the case where parallel flow $w = Uz$ on the z plane is mapped onto the ζ plane by mapping function $\zeta = z^{1/3}$.

13

Flow of a compressible fluid

Fluids have the capacity to change volume and density, i.e. compressibility. Gas is much more compressible than liquid.

Since liquid has low compressibility, when its motion is studied its density is normally regarded as unchangeable. However, where an extreme change in pressure occurs, such as in water hammer, compressibility is taken into account.

Gas has large compressibility but when its velocity is low compared with the sonic velocity the change in density is small and it is then treated as an incompressible fluid.

Nevertheless, when studying the atmosphere with large altitude changes, high-velocity gas flow in a pipe with large pressure difference, the drag sustained by a body moving with significant velocity in a calm gas, and the flow which accompanies combustion, etc., change of density must be taken into account.

As described later, the parameter expressing the degree of compressibility is the Mach number M. Supersonic flow, where $M > 1$, behaves very differently from subsonic flow where $M < 1$.

In this chapter, thermodynamic characteristics will be explained first, followed by the effects of sectional change in isentropic flow, flow through a convergent nozzle, and flow through a convergent–divergent nozzle. Then the adiabatic but irreversible shock wave will be explained, and finally adiabatic pipe flow with friction (Fanno flow) and pipe flow with heat transfer (Rayleigh flow).

13.1 Thermodynamical characteristics

Now, with the specific volume v and density ρ,

$$\rho v = 1 \tag{13.1}$$

A gas having the following relationship between absolute temperature T and pressure p

$$pv = RT \tag{13.2}$$

or

$$p = R\rho T \tag{13.3}$$

is called a perfect gas. Equations (13.2) and (13.3) are called its equations of state. Here R is the gas constant, and

$$R = \frac{R_0}{\mathscr{M}}$$

where R_0 is the universal gas constant ($R_0 = 8314\,\mathrm{J/(kg\,K)}$) and $\mathscr{M}$ is the molecular weight. For example, for air, assuming $\mathscr{M} = 28.96$, the gas constant is

$$R = \frac{8314}{28.96} = 287\,\mathrm{J/(kg\,K)} = 287\,\mathrm{m^2/(s^2\,K)}$$

Then, assuming internal energy and enthalpy per unit mass e and h respectively,

specific heat at constant volume: $c_v = \left(\dfrac{\partial e}{\partial T}\right)_v$ $de = c_v\,dT$ (13.4)

Specific heat at constant pressure: $c_p = \left(\dfrac{\partial h}{\partial T}\right)_p$ $dh = c_p\,dT$ (13.5)

Here

$$h = e + pv \tag{13.6}$$

According to the first law of thermodynamics, when a quantity of heat dq is supplied to a system, the internal energy of the system increases by de, and work $p\,dv$ is done by the system. In other words,

$$dq = de + p\,dv \tag{13.7}$$

From the equation of state (13.2),

$$p\,dv + v\,dp = R\,dT \tag{13.8}$$

From eqn (13.6),

$$dh = de + p\,dv + v\,dp \tag{13.9}$$

Now, since $dp = 0$ in the case of constant pressure change, eqns (13.8) and (13.9) become

$$p\,dv = R\,dT \tag{13.10}$$

$$dh = de + p\,dv = dq \tag{13.11}$$

Substitute eqns (13.4), (13.5), (13.10) and (13.11) into (13.7),

$$c_p\,dT = c_v\,dT + R\,dT$$

which becomes

$$c_p - c_v = R \tag{13.12}$$

Now, $c_p/c_v = k$ (k: ratio of specific heats (isentropic index)), so

$$c_p = \frac{k}{k-1} R \qquad (13.13)$$

$$c_v = \frac{1}{k-1} R \qquad (13.14)$$

Whenever heat energy dq is supplied to a substance of absolute temperature T, the change in entropy ds of the substance is defined by the following equation:

$$ds = dq/T \qquad (13.15)$$

As is clear from this equation, if a substance is heated the entropy increases, while if it is cooled the entropy decreases. Also, the higher the gas temperature, the greater the added quantity of heat for the small entropy increase.

Rewrite eqn (13.15) using eqns (13.1), (13.2), (13.12) and (13.13), and the following equation is obtained:[1]

$$\frac{dq}{T} = c_v \, d(\log pv^k) \qquad (13.16)$$

When changing from state (p_1, v_1) to state (p_2, v_2), if reversible, the change in entropy is as follows from eqns (13.15) and (13.16):

$$s_2 - s_1 = c_v \log\left(\frac{p_2 v_2^k}{p_1 v_1^k}\right) \qquad (13.17)$$

In addition, the relationships of eqns (13.18)–(13.20) are also obtained.[2]

---

[1] From

$$pv = RT \qquad \frac{dp}{p} + \frac{dv}{v} = \frac{dT}{T}$$

Therefore

$$\frac{dq}{T} = c_v \frac{dT}{T} + \frac{p}{T} dv = c_v \frac{dT}{T} + R \frac{dv}{v} = c_v \frac{dp}{p} + c_p \frac{dv}{v} = c_v \left(\frac{dp}{p} + k \frac{dv}{v}\right)$$

[2] Equations (13.18), (13.19) and (13.20) are respectively induced from the following equations:

$$ds = \frac{dq}{T} = c_v \frac{dT}{T} - R \frac{d\rho}{\rho} = c_v \frac{dT}{T} - (k-1)c_v \frac{d\rho}{\rho}$$

$$ds = \frac{dq}{T} = c_v \frac{dT}{T} + R \frac{dv}{v} = (c_v - R)\frac{dT}{T} + R \frac{dv}{v} = kc_v \frac{dT}{T} - (k-1)c_v \frac{dp}{p}$$

$$ds = \frac{dq}{T} = c_v \frac{dp}{p} + c_p \frac{dv}{v} = c_v \frac{dp}{p} - c_p \frac{d\rho}{\rho} = c_v \frac{dp}{p} - kc_v \frac{d\rho}{\rho}$$

$$S_2 - S_1 = c_v \log \left[\frac{T_2}{T_1} \left(\frac{\rho_1}{\rho_2} \right)^{k-1} \right] \tag{13.18}$$

$$S_2 - S_1 = c_v \log \left[\left(\frac{T_2}{T_1} \right)^k \left(\frac{p_1}{p_2} \right)^{k-1} \right] \tag{13.19}$$

$$S_2 - S_1 = c_v \log \left[\frac{p_2}{p_1} \left(\frac{\rho_1}{\rho_2} \right)^k \right] \tag{13.20}$$

for the reversible adiabatic (isentropic) change, $ds = 0$. Putting the proportional constant equal to c, eqn (13.17) gives (13.21), or eqn (13.22) from (13.20). That is,

$$pv^k = c \tag{13.21}$$

$$p = c\rho^k \tag{13.22}$$

Equations (13.18) and (13.19) give the following equation:

$$T = c\rho^{k-1} = cp^{(k-1)/k} \tag{13.23}$$

When a quantity of heat ΔQ transfers from a high-temperature gas at T_1 to a low-temperature gas at T_2, the changes in entropy of the respective gases are $-\Delta Q/T_1$ and $\Delta Q/T_2$. Also, the value of their sum is never negative.[3] Using entropy, the second law of thermodynamics could be expressed as 'Although the grand total of entropies in a closed system does not change if a reversible change develops therein, it increases if any irreversible change develops.' This is expressed by the following equation:

$$ds \geq 0 \tag{13.24}$$

Consequently, it can also be said that 'entropy in nature increases'.

13.2 Sonic velocity

It is well known that when a minute disturbance develops in a gas, the resulting change in pressure propagates in all directions as a compression wave (longitudinal wave, pressure wave), which we feel as a sound. Its propagation velocity is called the sonic velocity.

Here, for the sake of simplicity, assume a plane wave in a stationary fluid in a tube of uniform cross-sectional area A as shown in Fig. 13.1. Assume that, due to a disturbance, the velocity, pressure and density increase by u, dp and $d\rho$ respectively. Between the wavefront which has advanced at sonic velocity a and the starting plane is a section of length l where the pressure has increased. Since the wave travel time, during which the pressure increases in this section, is $t = l/a$, the mass in this section increases by $Al \, d\rho/t = Aa \, d\rho$

---

[3] In a reversible change where an ideal case is assumed, the heat shifts between gases of equal temperature. Therefore, $ds = 0$.

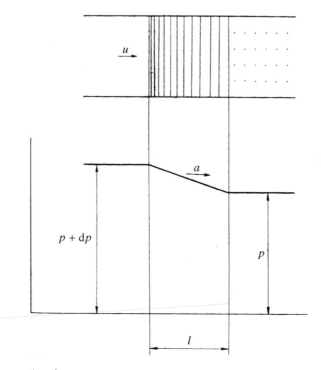

Fig. 13.1 Propagation of pressure wave

per unit time. In order to supplement it, gas of mass $Au(\rho + d\rho) = Au\rho$ flows in through the left plane. In other words, the continuity equation in this case is

$$Aa\,d\rho = Au\rho$$

or

$$a\,d\rho = u\rho \qquad (13.25)$$

The fluid velocity in this section changes from 0 to u in time t. In other words, the velocity can be regarded as having uniform acceleration $u/t = ua/l$. Taking its mass as $Al\rho$ and neglecting $d\rho$ in comparison with ρ, the equation of motion is

$$Al\rho\frac{ua}{l} = A\,dp$$

or

$$\rho au = dp \qquad (13.26)$$

Eliminate u in eqns (13.25) and (13.26), and

$$a = \sqrt{dp/d\rho} \qquad (13.27)$$

is obtained.

Since a sudden change in pressure is regarded as adiabatic, the following equation is obtained from eqns (13.3) and (13.23):[4]

$$a = \sqrt{kRT} \qquad (13.28)$$

In other words, the sonic velocity is proportional to the square root of absolute temperature. For example, for $k = 1.4$ and $R = 287\,\mathrm{m}^2/(\mathrm{s}^2\,\mathrm{K})$,

$$a = 20\sqrt{T} \quad (a = 340\,\mathrm{m/s} \text{ at } 16°\mathrm{C}\,(289\,\mathrm{K})) \qquad (13.29)$$

Next, if the bulk modulus of fluid is K, from eqns (2.13) and (2.15),

$$dp = -K\frac{dv}{v} = K\frac{d\rho}{\rho}$$

and

$$\frac{dp}{d\rho} = \frac{K}{\rho}$$

Therefore, eqn (13.27) can also be expressed as follows:

$$a = \sqrt{K/\rho} \qquad (13.30)$$

13.3 Mach number

The ratio of flow velocity u to sonic velocity a, i.e. $M = u/a$, is called Mach number (see Section 10.4.1). Now, consider a body placed in a uniform flow of velocity u. At the stagnation point, the pressure increases by $\Delta p = \rho U^2/2$ in approximation of eqn (9.1). This increased pressure brings about an increased density $\Delta\rho = \Delta p/a^2$ from eqn (13.27). Consequently,

$$M = \frac{U}{a} = \frac{1}{a}\sqrt{\frac{2\Delta p}{\rho}} = \sqrt{\frac{2\Delta\rho}{\rho}} \qquad (13.31)$$

In other words, the Mach number is a non-dimensional number expressing the compressive effect on the fluid. From this equation, the Mach number M corresponding to a density change of 5% is approximately 0.3. For this reason steady flow can be treated as incompressible flow up to around Mach number 0.3.

Now, consider the propagation of a sonic wave. This minute change in pressure, like a sound, propagates at sonic velocity a from the sonic source in all directions as shown in Fig. 13.2(a). A succession of sonic waves is produced cyclically from a sonic source placed in a parallel flow of velocity u. When u is smaller than a, as shown in Fig. 13.2(b), i.e. if $M < 1$, the wavefronts propagate at velocity $a - u$ upstream but at velocity $a + u$ downstream. Consequently, the interval between the wavefronts is dense upstream while being sparse

---

[4] $p = c\rho^k$, $dp/d\rho = ck\rho^{k-1} = kp/\rho = kRT$.

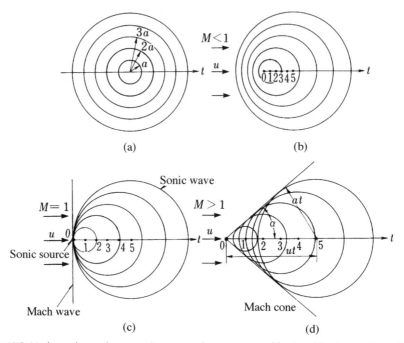

Fig. 13.2 Mach number and propagation range of a sonic wave: (a) calm; (b) subsonic ($M < 1$); (c) sonic ($M = 1$); (d) supersonic ($M > 1$)

downstream. When the upstream wavefronts therefore develop a higher frequency tone than those downstream this produces the Doppler effect.

When $u = a$, i.e. $M = 1$, the propagation velocity is just zero with the sound propagating downstream only. The wavefront is now as shown in Fig. 13.2(c), producing a Mach wave normal to the flow direction.

When $u > a$, i.e. $M > 1$, the wavefronts are quite unable to propagate upstream as in Fig. 13.2(d), but flow downstream one after another. The envelope of these wavefronts forms a Mach cone. The propagation of sound is limited to the inside of the cone only. If the included angle of this Mach cone is 2α, then[5]

$$\sin \alpha = a/u = 1/M \tag{13.32}$$

is called the Mach angle.

13.4 Basic equations for one-dimensional compressible flow

For a constant mass flow m of fluid density ρ flowing at velocity u through section area A, the continuity equation is

[5] Actually, the three-dimensional Mach line forms a cone, and the Mach angle is equal to its semi-angle.

$$m = \rho u A = \text{constant} \tag{13.33}$$

or by logarithmic differentiation

$$\frac{d\rho}{\rho} + \frac{du}{u} + \frac{dA}{A} = 0 \tag{13.34}$$

Euler's equation of motion in the steady state along a streamline is

$$\frac{1}{\rho}\frac{dp}{ds} + \frac{d}{ds}\left(\frac{1}{2}u^2\right) = 0$$

or

$$\int \frac{dp}{\rho} + \frac{1}{2}u^2 = \text{constant} \tag{13.35}$$

Assuming adiabatic conditions from $p = c\rho^k$,

$$\int \frac{dp}{\rho} = \int ck\rho^{k-2}\, d\rho = \frac{k}{k-1}\frac{p}{\rho} + \text{constant}$$

Substituting into eqn (13.35),

$$\frac{k}{k-1}\frac{p}{\rho} + \frac{1}{2}u^2 = \text{constant} \tag{13.36}$$

or

$$\frac{k}{k-1}RT + \frac{1}{2}u^2 = \text{constant} \tag{13.37}$$

Equations (13.36) and (13.37) correspond to Bernoulli's equation for an incompressible fluid.

If fluid discharges from a very large vessel, $u = u_o \approx 0$ (using subscript 0 for the state variables in the vessel), eqn (13.37) gives

$$\frac{k}{k-1}RT + \frac{1}{2}u^2 = \frac{k}{k-1}RT_0$$

or

$$\frac{T_0}{T} = 1 + \frac{1}{RT}\frac{k-1}{k}\frac{u^2}{2} = 1 + \frac{k-1}{2}M^2 \tag{13.38}$$

In this equation, T_0, T and $\dfrac{1}{R}\dfrac{k-1}{k}\dfrac{u^2}{2}$ are respectively called the total temperature, the static temperature and the dynamic temperature.

From eqns (13.23) and (13.38),

$$\frac{p_0}{p} = \left(\frac{T_0}{T}\right)^{k/(k-1)} = \left(1 + \frac{k-1}{2}M^2\right)^{k/(k-1)} \tag{13.39}$$

This is applicable to a body placed in the flow, e.g. between the stagnation point of a Pitot tube and the main flow.

Correction to a Pitot tube (see Section 11.1.1)

Putting p_∞ as the pressure at a point not affected by a body and making a binomial expansion of eqn (13.39), then (in the case where $M < 1$)

Table 13.1 Pitot tube correction

M	0	0.1	0.2	0.3	0.4	0.5	0.6	0.7	0.8
$(p_0 - p_\infty)/\frac{1}{2}\rho u^2 = c$	1.000	1.003	1.010	1.023	1.041	1.064	1.093	1.129	1.170
Relative error of $u = (\sqrt{c} - 1) \times 100\%$	0	0.15	0.50	1.14	2.03	3.15	4.55	6.25	8.17

$$p_0 = p_\infty \left(1 + \frac{k}{2} M^2 + \frac{k}{8} M^4 + \frac{k(2-k)}{48} M^6 + \Lambda \right)$$

$$= p_\infty + \frac{1}{2} \rho u^2 \left(1 + \frac{1}{4} M^2 + \frac{2-k}{24} M^4 + \Lambda \right)$$

(13.40)[6]

For an incompressible fluid, $p_0 = p_\infty + \frac{1}{2}\rho u^2$. Consequently, for the case when the compressibility of fluid is taken into account, the correction appearing in Table 13.1 is necessary.

From Table 13.1, it is found that, when $M = 0.7$, the true flow velocity is approximately 6% less than if the fluid was considered to be incompressible.

13.5 Isentropic flow

13.5.1 Flow in a pipe (Effect of sectional change)

Consider the flow in a pipe with a gradual sectional change, as shown in Fig. 13.3, having its properties constant across any section. For the fluid at sections 1 and 2 in Fig. 13.3,

continuity equation: $\qquad \dfrac{d\rho}{\rho} + \dfrac{du}{u} + \dfrac{dA}{A} = 0$ (13.41)

equation of momentum conservation: $\qquad - dp\, A = (A\rho u)du$ (13.42)

isentropic relationship: $\qquad p = c\rho^k$ (13.43)

sonic velocity: $\qquad a^2 = \dfrac{dp}{d\rho}$ (13.44)

From eqns (13.41), (13.42) and (13.44),

$$- a^2 \, d\rho = \rho u \, du = \rho u^2 \frac{du}{u}$$

$$M^2 \frac{du}{u} = -\frac{d\rho}{\rho} = \frac{du}{u} + \frac{dA}{A}$$

$$p_\infty k M^2 = p_\infty k \frac{u^2}{a^2} = p_\infty \frac{ku^2}{kRT} = \frac{p_\infty}{RT} u^2 = \rho u^2$$

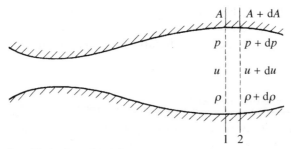

Fig. 13.3 Flow in pipe with gentle sectional change

Therefore

$$(M^2 - 1)\frac{du}{u} = \frac{dA}{A} \tag{13.45}$$

or

$$\frac{du}{dA} = \frac{1}{M^2 - 1}\frac{u}{A} \tag{13.46}$$

Also,

$$\frac{d\rho}{\rho} = -M^2\frac{du}{u} \tag{13.47}$$

Therefore,

$$-\frac{d\rho}{\rho}\Big/\frac{du}{u} = M^2 \tag{13.48}$$

From eqn (13.46), when $M < 1$, $du/dA < 0$, i.e. the flow velocity decreases with increased sectional area, but when $M > 1$, $-d\rho/\rho > du/u$, i.e. for supersonic flow the density decreases at a faster rate than the velocity increases. Consequently, for mass continuity, the surprising fact emerges that in order to increase the flow velocity the section area should increase rather than decrease, as for subsonic flow.

Table 13.2 Subsonic flow and supersonic flow in one-dimensional isentropic flow

	Flow state			
	Subsonic		Supersonic	
Changing item				
Changing area	−	+	−	+
Changing velocity/Mach number	+	−	−	+
Changing density/pressure/temperature	−	+	+	−

From eqn (13.47), the change in density is in reverse relationship to the velocity. Also from eqn (13.23), the pressure and the temperature change in a similar manner to the density. The above results are summarised in Table 13.2.

13.5.2 Convergent nozzle

Gas of pressure p_0, density ρ_0 and temperature T_0 flows from a large vessel through a convergent nozzle into the open air of back pressure p_b isentropically at velocity u, as shown in Fig. 13.4. Putting p as the outer plane pressure, from eqn (13.36)

$$\frac{u^2}{2} + \frac{k}{k-1}\frac{p}{\rho} = \frac{k}{k-1}\frac{p_0}{\rho_0}$$

Using eqn (13.23) with the above equation,

$$u = \sqrt{2\frac{k}{k-1}\frac{p_0}{\rho_0}\left[1 - \left(\frac{p}{p_0}\right)^{(k-1)/k}\right]} \qquad (13.49)$$

Therefore, the flow rate is

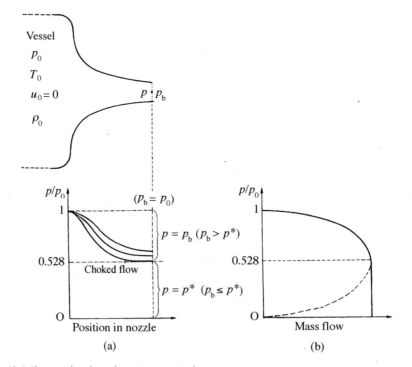

Fig. 13.4 Flow passing through convergent nozzle

$$m = \rho u A = A \sqrt{ 2 \frac{k}{k-1} p_0 \rho_0 \left(\frac{p}{p_0} \right)^{2/k} \left[1 - \left(\frac{p}{p_0} \right)^{(k-1)/k} \right] } \qquad (13.50)$$

Writing $p/p_0 = x$, then

$$\frac{\partial m}{\partial x} = 0 \quad x = \frac{p}{p_0} = \left(\frac{2}{k+1} \right)^{k/(k-1)} \qquad (13.51)$$

When p/p_0 has the value of eqn (13.51), m is maximum. The corresponding pressure is called the critical pressure and is written as p^*. For air,

$$p^*/p_0 = 0.528 \qquad (13.52)$$

Using the relationship between m and p/p_0 in eqn (13.50), the maximum flow rate occurs when $p/p_0 = 0.528$ as shown in Fig 13.4(b). Thereafter, however much the pressure p_b downstream is lowered, the pressure there cannot propagate towards the nozzle because it is discharging at sonic velocity. Therefore, the pressure of the air in the outlet plane remains p^*, and the mass flow rate does not change. In this state the flow is called choked.

Substitute eqn (13.51) into (13.49) and use the relationship $p_0/\rho_0^k = p/\rho^k$ to obtain

$$u^* = \sqrt{ k \frac{p}{\rho} } = a \qquad (13.53)$$

In other words, for $M = 1$, under these conditions u is called the critical velocity and is written as u^*. At the same time

$$\frac{\rho^*}{\rho_0} = \left(\frac{2}{k+1} \right)^{1/(k-1)} = 0.634 \qquad (13.54)$$

$$\frac{T^*}{T_0} = \frac{2}{k+1} = 0.833 \qquad (13.55)$$

The relationships of the above equations (13.52), (13.54) and (13.55) show that, at the critical outlet state $M = 1$, the critical pressure falls to 52.5% of the pressure in the vessel, while the critical density and the critical temperature respectively decrease by 37% and 17% from those of the vessel.

13.5.3 Convergent–divergent nozzle

A convergent–divergent nozzle (also called the de Laval nozzle) is, as shown in Fig. 13.5,[7] a convergent nozzle followed by a divergent length. When back pressure p_b outside the nozzle is reduced below p_0, flow is established. So long as the fluid flows out through the throat section without reaching the critical pressure the general behaviour is the same as for incompressible fluid.

When the back pressure decreases further, the pressure at the throat section

[7] Liepmann, H. W. and Roshko, A., *Elements of Gasdynamics*, (1975), 127, John Wiley, New York.

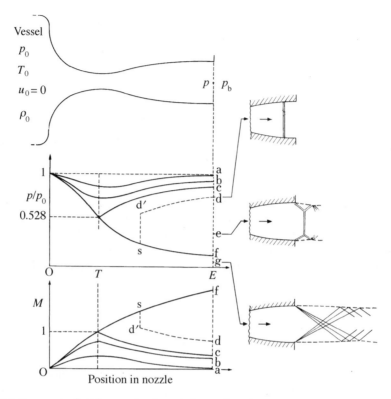

Fig. 13.5 Compressive fluid flow passing through divergent nozzle

reaches the critical pressure and $M = 1$; thereafter the flow in the divergent port is at least initially supersonic. However, unless the back pressure is low enough, supersonic velocity cannot be maintained. Instead, a shock wave develops, after which the flow becomes subsonic. As the back pressure is replaced, the shock moves further away from the diverging length to the exit plane and eventually disappears, giving a perfect expansion.

A real ratio A/A^* between the outlet section and the throat giving this perfect expansion is called the area ratio, and, using eqns (13.50) and (13.51),

$$\frac{A}{A^*} = \left(\frac{2}{k+1}\right)^{1/(k-1)} \left(\frac{p_0}{p}\right)^{1/k} \Big/ \sqrt{\frac{k+1}{k-1}\left[1 - \left(\frac{p_0}{p}\right)^{(1-k)/k}\right]} \tag{13.56}$$

13.6 Shock waves

When air undergoes large and rapid compression (e.g. following an explosion, the release of engine gases into an exhaust pipe, or where an aircraft or a bullet flies at supersonic velocity) a thin wave of large pressure

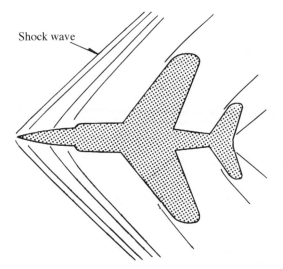

Fig. 13.6 Jet plane flying at supersonic velocity

change is produced as shown in Figs 13.6 and 13.7. Since the state of gas changes adiabatically, an increased temperature accompanies this increased pressure. As shown in Fig. 13.8(a), the wave face at the rear of the compression wave, being at a higher temperature, propagates faster than the wave face at the front. The rear therefore gradually catches up with the front until finally, as shown in Fig. 13.8(b), the wave faces combine into a thin wave increasing the pressure discontinuously. Such a pressure discontinuity is called a shock wave, which is only associated with an increase, rather than a reduction, in pressure in the flow direction.

Since a shock wave is essentially different from a sound wave because of the large change in pressure, the propagation velocity of the shock is larger, and the larger the pressure rise, the greater the propagation velocity.

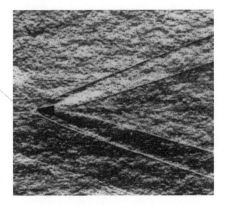
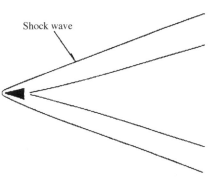

Fig. 13.7 Cone flying at supersonic velocity (Schlieren method) in air, Mach 3

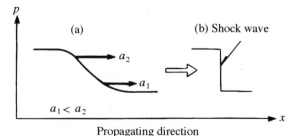

Fig. 13.8 Propagation of a compression wave

If a long cylinder is partitioned with Cellophane film or aluminium foil to give a pressure difference between the two sections, and then the partition is ruptured, a shock wave develops. The shock wave in this case is at right angles to the flow, and is called a normal shock wave. The device itself is called a shock tube.

As shown in Fig. 13.9, the states upstream and downstream of the shock wave are respectively represented by subscripts 1 and 2. A shock wave Δx is so thin, approximately micrometres at thickest, that it is normally regarded as having no thickness.

Now, assuming $A_1 = A_2$, the continuity equation is

$$\rho_1 u_1 = \rho_2 u_2 \tag{13.57}$$

the equation of momentum conservation is

$$p_1 + \rho_1 u_1^2 = p_2 + \rho_2 u_2^2 \tag{13.58}$$

and the equation of energy conservation is

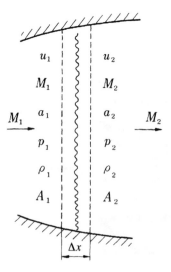

Fig. 13.9 Normal shock wave

$$\frac{u_1^2}{2} + \frac{k}{k-1}\frac{p_1}{\rho_1} = \frac{u_2^2}{2} + \frac{k}{k-1}\frac{p_2}{\rho_2}$$

or

$$u_1^2 - u_2^2 = \frac{2k}{k-1}\left(\frac{p_2}{\rho_2} - \frac{p_1}{\rho_1}\right) \tag{13.59}$$

From eqns (13.57) and (13.58),

$$u_1^2 = \frac{p_2 - p_1}{\rho_2 - \rho_1}\frac{\rho_2}{\rho_1} \tag{13.60}$$

$$u_2^2 = \frac{p_2 - p_1}{\rho_2 - \rho_1}\frac{\rho_1}{\rho_2} \tag{13.61}$$

Substituting eqns (13.60) and (13.61) into (13.59),

$$\frac{\rho_2}{\rho_1} = \frac{[(k+1)/(k-1)](p_2/p_1) + 1}{[(k+1)/(k-1)] + p_2/p_1} = \frac{u_1}{u_2} \tag{13.62}$$

Or, using eqn (13.3),

$$\frac{T_2}{T_1} = \frac{[(k+1)/(k-1)] + p_2/p_1}{[(k+1)/(k-1)] + p_1/p_2} \tag{13.63}$$

Equations (13.62) and (13.63), which are called the Rankine–Hugoniot equations, show the relationships between the pressure, density and temperature ahead of and behind a shock wave. From the change of entropy associated with these equations it can be deduced that a shock wave develops only when the upstream flow is supersonic.[8]

It has already been explained that when a supersonic flow strikes a particle, a Mach line develops. On the other hand, when a supersonic flow flows along a plane wall, numerous parallel Mach lines develop as shown in Fig. 13.10(a).

When supersonic flow expands around a curved wall as shown in Fig. 13.10(b), the Mach waves rotate, forming an expansion 'fan'. This flow is called a Prandtl–Meyer expansion.

In Fig. 13.10(c), a compressive supersonic flow develops where numerous Mach lines change their direction, converging and overlapping to develop a sharp change of pressure and density, i.e. a shock wave.

---

[8] From eqns (13.57) and (13.58),

$$\frac{p_2}{p_1} = 1 + \frac{2k}{k+1}(M_1^2 - 1)$$

Likewise

$$\frac{p_1}{p_2} = 1 + \frac{2k}{k+1}(M_2^2 - 1)$$

Therefore

$$M_2^2 = \frac{2 + (k-1)M_1^2}{2kM_1^2 - (k-1)}$$

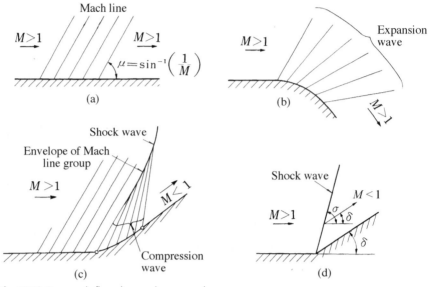

Fig. 13.10 Supersonic flow along various wave shapes

Figure 13.10(d) shows the ultimate state of a shock wave due to supersonic flow passing along this concave wall. Here, δ is the deflection angle and σ is the shock wave angle.

A shock wave is called a normal shock wave when $\sigma = 90°$ and an oblique shock wave in other cases.

From Fig. 13.11, the following relationships arise between the normal component u_n and the tangential component u_t of the flow velocity through an oblique shock wave:

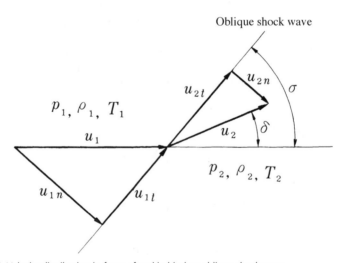

Fig. 13.11 Velocity distribution in front of and behind an oblique shock wave

$$u_{1n} = u_1 \sin \sigma \qquad u_{1t} = u_1 \cos \sigma$$
$$u_{2n} = u_2 \sin(\sigma - \delta) \quad u_{2t} = u_2 \cos(\sigma - \delta) \tag{13.64}$$

From the momentum equation in the tangential direction, since there is no pressure gradient,

$$u_{1t} = u_{2t} \tag{13.65}$$

From the momentum equation in the normal direction,

$$u_{1n}^2 - u_{2n}^2 = \frac{2k}{k-1}\left(\frac{p_2}{\rho_2} - \frac{p_1}{\rho_1}\right) \tag{13.66}$$

This equation is in the same form as eqn (13.59), and the Rankine–Hugoniot equations apply. When combined with eqn (13.64), the following relationship is developed between δ and σ:

$$\cos \delta = \left(\frac{k+1}{2}\frac{M_1^2}{M_1^2 \sin^2 \sigma - 1} - 1\right) \tan \sigma \tag{13.67}$$

When the shock angle $\sigma = 90°$ and $\sigma = \sin^{-1}(1/M_1)$, $\delta = 0$ so the maximum value δ_m of δ must lie between these values.

The shock wave in the case of a body where $\delta < \delta_m$ (Fig. 13.12(a)) is attached to the sharp nose A. In the case of a body where $\delta > \delta_m$ (Fig. 13.12(b)), however, the shock wave detaches and stands off from nose A.

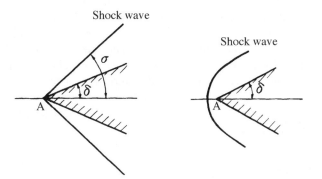

Fig. 13.12 Flow pattern and shock wave around body placed in supersonic flow: (a) shock wave attached to wedge; (b) detached shock wave

13.7 Fanno flow and Rayleigh flow

Since an actual flow of compressible fluid in pipe lines and similar conduits is always affected by the friction between the fixed wall and the fluid, it can be adiabatic but not isentropic. Such an adiabatic but irreversible (i.e. non-isentropic) flow is called Fanno flow.

Alternatively, in a system of flow forming a heat exchanger or combustion process, friction may be neglected but transfer of heat must be taken into

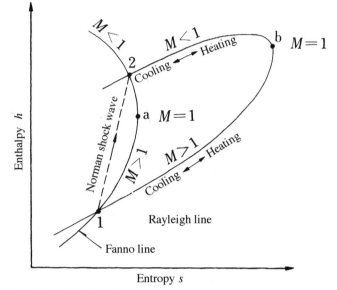

Fig. 13.13 Fanno line and Rayleigh line

account. Such a flow without friction through a pipe with heat transmission is called Rayleigh flow.

Figure 13.13 shows a diagram of both of these flows in a pipe with fixed section area. The lines appearing there are called the Fanno line and Rayleigh line respectively. For both of them, points a or b of maximum entropy correspond to the sonic state $M = 1$. The curve above these points corresponds to subsonic velocity and that below to supersonic velocity.

The states immediately ahead of and behind the normal shock wave are expressed by the intersection points 1 and 2 of these two curves. For the flow through the shock wave, only the direction of increased entropy, i.e. the discontinuous change, $1 \rightarrow 2$ is possible.

13.8 Problems

1. When air is regarded as a perfect gas, what is the density in kg/m^3 of air at 15°C and 760 mm Hg?

2. Find the velocity of sound propagating in hydrogen at 16°C.

3. When the velocity is 30 m/s, pressure 3.5×10^5 Pa and temperature 150°C at a point on a streamline in an isentropic air flow, obtain the pressure and temperature at the point on the same streamline of velocity 100 m/s.

4. Find the temperature, pressure and density at the front edge (stagnation point) of a wing of an aircraft flying at 900 km/h in calm air of pressure 4.5×10^4 Pa and temperature −26°C.

5. From a Schlieren photograph of a small bullet flying in air at 15°C and standard atmospheric pressure, it was noticed that the Mach angle was 50°. Find the velocity of this bullet.

6. When a Pitot tube was inserted into an air flow at high velocity, the pressure at the stagnation point was 1×10^5 Pa, the static pressure was 7×10^4 Pa, and the air temperature was $-10°C$. Find the velocity of this air flow.

7. Air of gauge pressure 6×10^4 Pa and temperature 20°C is stored in a large tank. When this air is released through a convergent nozzle into air of 760 mm Hg, find the flow velocity at the nozzle exit.

8. Air of gauge pressure 1.2×10^5 Pa and temperature 15°C is stored in a large tank. When this air is released through a convergent nozzle of exit area 3 cm^2 into air of 760 mm Hg, what is the mass flow?

9. Find the divergence ratio necessary for perfectly expanding air under standard conditions down to 100 mm Hg absolute pressure through a convergent–divergent nozzle.

10. The nozzle for propelling a rocket is a convergent–divergent nozzle of throat cross-sectional area 500 cm^2. Regard the combustion gas as a perfect gas of mean molecular weight 25.8 and $\kappa = 1.25$. In order to make the combustion gas of pressure 32×10^5 Pa and temperature 3300 K expand perfectly out from the combustion chamber into air of 1×10^5 Pa, what should be the cross-sectional area at the nozzle exit?

11. When the rocket in Problem 10 flies at an altitude where the pressure is 2×10^4 Pa, what is the obtainable thrust from the rocket?

12. A supersonic flow of Mach 2, pressure 5×10^4 Pa and temperature $-15°C$ develops a normal shock wave. What is the Mach number, flow velocity and pressure behind the wave?

Unsteady Flow

From olden times fluid had mostly been utilised mechanically for generating motive power, but recently it has been utilised for transmitting or automatically controlling power too. High-pressure fluid has to be used in these systems for high speed and good response. Consequently the issue of unsteady flow has become very important.

14.1 Vibration of liquid column in a U-tube

When the viscous frictional resistance is zero, by Newton's laws (Fig. 14.1),

$$\rho g(z_2 - z_1)A = -\rho A l \frac{dv}{dt} \tag{14.1}$$

$$g(z_2 - z_1) + l \frac{dv}{dt} = 0 \tag{14.2}$$

Moving the datum for height to the balanced state position, then

$$g(z_2 - z_1) = 2gz$$

Also,

$$\frac{dv}{dt} = \frac{d^2z}{dt^2}$$

and from above,

$$\frac{d^2z}{dt^2} = -\frac{2g}{l}z \tag{14.3}$$

Therefore

$$z = C_1 \cos\sqrt{\frac{2g}{l}}t + C_2 \sin\sqrt{\frac{2g}{l}}t \tag{14.4}$$

Assuming the initial conditions are $t = 0$ and $z = z_0$, then $dz/dt = 0$, $C_1 = z_0$, $C_2 = 0$. Therefore

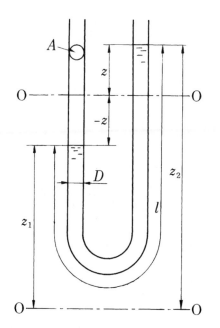

Fig. 14.1 Vibration of liquid column in a U-tube

$$z = z_0 \cos \sqrt{\frac{2g}{l}} t \qquad (14.5)$$

This means that the liquid surface makes a singular vibration of cycle $T = 2\pi\sqrt{l/2g}$.

14.1.2 Laminar frictional resistance

In this case, with the viscous frictional resistance in eqn (14.1),

$$g(z_2 - z_1) + l\frac{dv}{dt} + \frac{32vvl}{D^2} = 0 \qquad (14.6)$$

Substituting $2z = (z_2 - z_1)$ as above,

$$\frac{d^2z}{dt^2} + \frac{32v}{D^2}\frac{dz}{dt} + \frac{2g}{l}z = 0$$

$$\frac{d^2z}{dt^2} + 2\zeta\omega_n\frac{dz}{dt} + \omega_n^2 z = 0 \quad \text{where } \omega_n = \sqrt{\frac{2g}{l}} \text{ and } \zeta = \frac{16v}{D^2}\frac{1}{\omega_n} \qquad (14.7)$$

The general solution of eqn (14.7) is as follows:
  (a) when $\zeta < 1$

$$z = e^{-\zeta\omega_n t}\left[C_1 \sin\left(\omega_n\sqrt{1-\zeta^2}t\right) + C_2 \cos\left(\omega_n\sqrt{1-\zeta^2}t\right)\right] \qquad (14.8)$$

Assume $z = z_0$ and $dz/dt = 0$ when $t = 0$. Then

$$z = z_0 e^{-\zeta\omega_n t} \left[\frac{\zeta}{\sqrt{1 - \zeta^2}} \sin\left(\omega_n \sqrt{1 - \zeta^2}t\right) + \cos\left(\omega_n \sqrt{1 - \zeta^2}t\right) \right]$$

$$= \frac{z_0}{\sqrt{1 - \zeta^2}} e^{-\zeta\omega_n t} \sin\left(\omega_n \sqrt{1 - \zeta^2}t + \phi\right) \qquad (14.9)$$

$$\phi = \tan^{-1}\left(\frac{\sqrt{1 - \zeta^2}}{\zeta}\right)$$

(b) when $\zeta > 1$

$$z = z_0 e^{-\zeta\omega_n t} \left[\frac{\zeta}{\sqrt{\zeta^2 - 1}} \sinh\left(\omega_n \sqrt{\zeta^2 - 1}t\right) + \cosh\left(\omega_n \sqrt{\zeta^2 - 1}t\right) \right]$$

$$= \frac{z_0}{\sqrt{\zeta^2 - 1}} e^{-\zeta\omega_n t} \sinh\left(\omega_n \sqrt{\zeta^2 - 1}t + \phi\right) \qquad (14.10)$$

$$\phi = \tanh^{-1}\left(\frac{\sqrt{\zeta^2 - 1}}{\zeta}\right)$$

Equations (14.9) and (14.10) can be plotted using the non-dimensional quantities of $\omega_n t$, z/z_0 as shown in Fig. 14.2. With large frictional resistance there is no oscillation but as it becomes smaller a damped oscillation occurs.

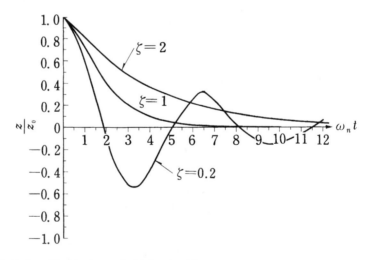

Fig. 14.2 Motion of liquid column with frictional resistance

14.2 Propagation of pressure in a pipe line

In the system shown in Fig. 14.3, a tank (capacity V) is connected to a pipe line (diameter D, section area A and length l). If the inlet pressure is suddenly

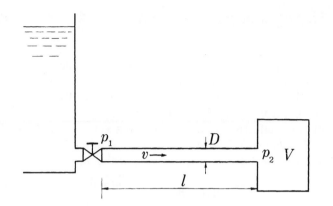

Fig. 14.3 System comprising pipe line and tank

changed (from 0 to p_1, say), it is desirable to know the response of the outlet pressure p_2. Assuming a pressure loss Δp due to tube friction, with instantaneous flow velocity v the equation of motion is

$$\rho A l = \frac{dv}{dt} = A(p_1 - p_2 - \Delta p) \qquad (14.11)$$

If v is within the range of laminar flow, then

$$\Delta p = \frac{32\mu l}{D^2} v \qquad (14.12)$$

Taking only the fluid compressibility β into account since the pipe is a rigid body,

$$dp_2 = \frac{1}{\beta} \frac{A v\, dt}{V} \qquad (14.13)$$

Substituting eqns (14.12) and (14.13) into (14.11) gives

$$\frac{d^2 p_2}{dt^2} + \frac{32v}{D^2} \frac{dp_2}{dt} + \frac{A}{\rho l \beta V}(p_2 - p_1) = 0$$

Now, writing

$$\omega_n = \sqrt{\frac{A}{\rho l \beta V}} \quad \zeta = \frac{16v}{D^2} \frac{1}{\omega_n} \quad p_2 - p_1 = p$$

then

$$\frac{d^2 p}{dt^2} + 2\zeta \omega_n \frac{dp}{dt} + \omega_n^2 p = 0 \qquad (14.14)$$

Since eqn (14.14) has the same form as eqn (14.7), the solution also has the same form as eqn (14.9) with the response tendency being similar to that shown in Fig. 14.2.

14.3 Transitional change in flow quantity in a pipe line

When the valve at the end of a pipe line of length l as shown in Fig. 14.4 is instantaneously opened, there is a time lapse before the flow reaches steady state. When the valve first opens, the whole of head H is used for accelerating the flow. As the velocity increases, however, the head used for acceleration decreases owing to the fluid friction loss h_1 and discharge energy h_2. Consequently, the effective head available to accelerate the liquid in the pipe becomes $\rho g(H - h_1 - h_2)$. So the equation of motion of the liquid in the pipe is as follows, putting A as the sectional area of the pipe,

$$\rho g A(H - h_1 - h_2) = \frac{\rho g A l}{g}\frac{dv}{dt} \tag{14.15}$$

giving

$$h_1 = \lambda\frac{l}{d}\frac{v^2}{2g} = k\frac{v^2}{2g} \quad h_2 = \frac{v^2}{2g}$$

and

$$H - (k+1)\frac{v^2}{2g} = \frac{l}{g}\frac{dv}{dt} \tag{14.16}$$

Assume that velocity v becomes v_0 (terminal velocity) in the steady state ($dv/dt = 0$). Then

$$(k+1)v_0^2 = 2gh$$

$$k = \frac{2gh}{v_0^2} - 1$$

Substituting the value of k above into eqn (14.16),

$$H\left(1 - \frac{v^2}{v_0^2}\right) = \frac{l}{g}\frac{dv}{dt}$$

$$dt = \frac{l}{gH}\frac{v_0^2}{v_0^2 - v^2}dv$$

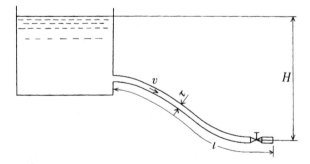

Fig. 14.4 Transient flow in a pipe

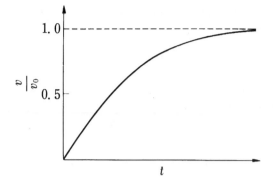

Fig. 14.5 Development of steady flow

or

$$t = \frac{lv_0}{2gH}\log\left(\frac{v_0 + v}{v_0 - v}\right) \tag{14.17}$$

Thus, time t for the flow to become steady is obtainable (Fig. 14.5).

Now, calculating the time from $v/v_0 = 0$ to $v/v_0 = 0.99$, the following equation can be obtained:

$$t = \frac{lv_0}{2gh}\log\left(\frac{1.99}{0.01}\right) = 2.646\frac{lv_0}{gh} \tag{14.18}$$

14.4 Velocity of pressure wave in a pipe line

The velocity of a pressure wave depends on the bulk modulus K (eqn (13.30)). The bulk modulus expresses the relationship between change of pressure on a fluid and the corresponding change in its volume. When a small volume V of fluid in a short length of rigid pipe experiences a pressure wave, the resulting reduction in volume dV_1 produces a reduction in length. If the pipe is elastic, however, it will experience radial expansion causing an increase in volume dV_2. This produces a further reduction in the length of volume V. Therefore, to the wave, the fluid appears more compressible, i.e. to have a lower bulk modulus. A modified bulk modulus K' is thus required which incorporates both effects.

From the definition equation (2.10),

$$-\frac{dp}{K} = \frac{dV_1}{V} \tag{14.19}$$

where the minus sign was introduced solely for the convenience of having positive values of K. Similarly, for positive K',

$$-\frac{dp}{K'} = \frac{dV_1 - dV_2}{V} \tag{14.20}$$

where the negative dV_2 indicates that, despite being a volume increase, it produces the equivalent effect of a volume reduction dV_1. Thus

$$\frac{1}{K'} = \frac{1}{K} + \frac{dV_2}{V\,dp} \tag{14.21}$$

If the elastic modulus (Young's modulus) of a pipe of inside diameter D and thickness b is E, the stress increase $d\sigma$ is

$$d\sigma = E\frac{dD}{D}$$

This hoop stress in the wall balances the internal pressure dp,

$$d\sigma = \frac{D}{2b}dp$$

Therefore

$$\frac{dD}{D} = \frac{D\,dp}{2bE}$$

Since $V = \pi D^2/4$ and $dV_2 = \pi D\,dD/2$ per unit length,

$$\frac{dV_2}{V} = 2\frac{dD}{D} = \frac{D\,dP}{bE} \tag{14.22}$$

Substituting eqn (14.22) into (14.21), then

$$\frac{1}{K'} = \frac{1}{K} + \frac{D}{bE}$$

or

$$K' = \frac{K}{1 + (D/b)(K/E)} \tag{14.23}$$

The sonic velocity a_0 in the fluid is, from eqn (13.30),

$$a_0 = \sqrt{K/\rho}$$

Therefore, the propagation velocity of the pressure wave in an elastic pipe is

$$a = \sqrt{\frac{K'}{\rho}} = \sqrt{\frac{K/\rho}{1 + (D/b)(K/E)}} = a_0\sqrt{\frac{1}{1 + (D/b)(K/E)}} \tag{14.24}$$

Since the values of D for steel, cast iron and concrete are respectively 206, 92.1 and 20.6 GPa, a is in the range 600–1200 m/s in an ordinary water pipe line.

14.5 Water hammer

Water flows in a pipe as shown in Fig. 14.6. If the valve at the end of the pipe is suddenly closed, the velocity of the fluid will abruptly decrease causing a mechanical impulse to the pipe due to a sudden increase in pressure of the fluid. Such a phenomenon is called water hammer. This phenomenon poses a

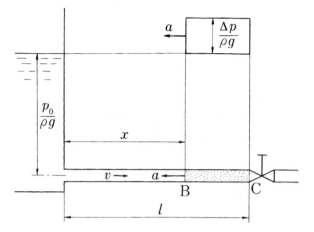

Fig. 14.6 Water hammer

very important problem in cases where, for example, a valve is closed to reduce the water flow in a hydraulic power station when the load on the water wheel is reduced. In general, water hammer is a phenomenon which is always possible whenever a valve is closed in a system where liquid is flowing.

14.5.1 Case of instantaneous valve closure

When the valve at pipe end C in Fig. 14.6 is instantaneously closed, the flow velocity v of the fluid in the pipe, and therefore also its momentum, becomes zero. Therefore, the pressure increases by dp. Since the following portions of fluid are also stopped one after another, dp propagates upstream. The propagation velocity of this pressure wave is expressed by eqn (14.24).

Given that an impulse is equal to the change of momentum,

$$dpA\frac{l}{a} = \rho Alv$$

or

$$dp = \rho va \qquad (14.25)$$

When this pressure wave reaches the pipe inlet, the pressurised pipe begins to discharge backwards into the tank at velocity v. The pressure reverts to the original tank pressure p_0, and the pipe, too, begins to contract to its original state. The low pressure and pipe contraction proceed from the tank end towards the valve at velocity a with the fluid behind the wave flowing at velocity v. In time $2l/a$ from the valve closing, the wave reaches the valve. The pressure in the pipe has reverted to the original pressure, with the fluid in the pipe flowing at velocity v. Since the valve is closed, however, the velocity there must be zero. This requires a flow at velocity $-v$ to propagate from the valve. This outflow causes the pressure to fall by dp. This $-dp$ propagates

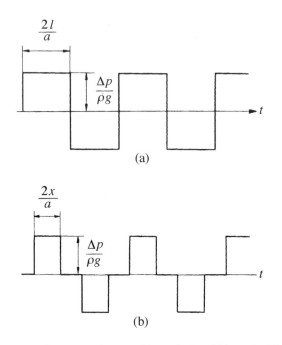

Fig. 14.7 Change in pressure due to water hammer, (a) at point C and (b) at point B in Fig. 14.6

upstream at velocity a. At time $3l/a$, from the valve closing, the liquid in the pipe is at rest with a uniform low pressure of $-dp$. Then, once again, the fluid flows into the low pressure pipe from the tank at velocity v and pressure p. The wave propagates downstream at velocity a. When it reaches the valve, the pressure in the pipe has reverted to the original pressure and the velocity to its original value. In other words, at time $4l/a$ the pipe reverts to the state when the valve was originally closed. The changes in pressure at points C and B in Fig. 14.6 are as shown in Fig. 14.7(a) and (b) respectively. The pipe wall around the pressurised liquid also expands, so that the waves propagate at velocity a as shown in eqn (14.24).

14.5.2 Valve closure in time t_c

When the valve closing time t_c is less than time $2l/a$ for the wave round-trip of the pipe line, the maximum pressure increase when the valve is closed is equal to that in eqn (14.25).

When the valve closing time t_c is longer than time $2l/a$, it is called slow closing, to which Allievi's equation applies (named after L. Allievi (1856–1941), Italian hydraulics scholar). That is,

$$\frac{p_{max}}{p_0} = 1 + \tfrac{1}{2}(n^2 + n\sqrt{n^2 + 4})$$ (14.26)

Here, p_{max} is the highest pressure generated when the valve is closed, p_0 is the pressure in the pipe when the valve is open, v is the flow velocity when the valve is open, and $n = \rho l v/(p_0 t_c)$. This equation does not account for pipe friction and the valve is assumed to be uniformly closed.

In practice, however, there is pipe friction and valve leakage occurs. To obtain such changes in the flow velocity or pressure, either graphical analysis[1] or computer analysis (see Section 15.1) using the method of characteristics may be used.

14.6 Problems

1. As shown in Fig. 14.8, a liquid column of length 1.225 m in a U-shaped pipe is allowed to oscillate freely. Given that at $t = 0$, $z = z_0 = 0.4$ m and $dz/dt = 0$, obtain

   (a) the velocity of the liquid column when $z = 0.2$ m, and
   (b) the oscillation cycle time.

   Ignore frictional resistance.

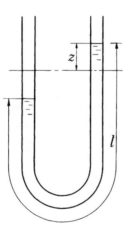

Fig. 14.8

2. Obtain the cycle time for the oscillation of liquid in a U-shaped tube whose arms are both oblique (Fig. 14.9). Ignore frictional resistance.

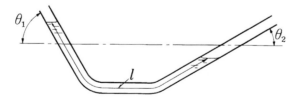

Fig. 14.9

---

[1] Parmakian, J., *Waterhammer Analysis*, (1963), 2nd edition, Dover, New York.

3. Oil of viscosity $v = 3 \times 10^{-5}\,\mathrm{m^2/s}$ extends over a 3 m length of a tube of diameter 2.5 cm, as shown in Fig. 14.8. Air pressure in one arm of the U-tube, which produces 40 cm of liquid column difference, is suddenly released causing the liquid column to oscillate. What is the maximum velocity of the liquid column if laminar frictional resistance occurs?

4. As shown in Fig. 14.10, a pipe line of diameter 2 m and length 400 m is connected to a tank of head 18 m. Find the time from the sudden opening of the valve for the exit velocity to reach 90% of the final velocity. Use a friction coefficient for the pipe of 0.03.

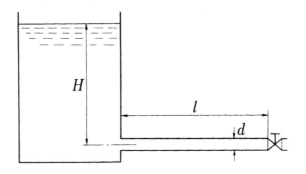

Fig. 14.10

5. Find the velocity of a pressure wave propagating in a water-filled steel pipe of inside diameter 2 cm and wall thickness 1 cm, if the bulk modulus $K = 2.1 \times 10^9\,\mathrm{Pa}$, density $\rho = 1000\,\mathrm{kg/m^3}$ and Young's modulus for steel $E = 2.1 \times 10^{11}\,\mathrm{Pa}$.

6. Water flows at a velocity of 3 m/s in the steel pipe in Problem 5, of length 1000 m. Obtain the increase in pressure when the valve is shut instantaneously.

7. The steady-state pressure of water flowing in the pipe line in Problem 6, at a velocity of 3 m/s, is $5 \times 10^5\,\mathrm{Pa}$. What is the maximum pressure reached when the valve is shut in 5 seconds?

15

Computational fluid dynamics

For the flow of an incompressible fluid, if the Navier–Stokes equations of motion and the continuity equation are solved simultaneously under given boundary conditions, an exact solution should be obtained. However, since the Navier–Stokes equations are non-linear, it is difficult to solve them analytically.

Nevertheless, approximate solutions are obtainable, e.g. by omitting the inertia terms for a flow whose Re is small, such as slow flow around a sphere or the flow of an oil film in a sliding bearing, or alternatively by neglecting the viscosity term for a flow whose Re is large, such as a fast free-stream flow around a wing. But for intermediate Re, the equations cannot be simplified because the inertia term is roughly as large as the viscosity term. Consequently there is no other way than to obtain the approximate solution numerically.

For a compressible fluid, it is further necessary to solve the equation of state and the energy equation simultaneously with respect to the thermodynamical properties. Thus, multi-dimensional shock wave problems can only be solved by relying upon numerical solution methods.

Of late, with the progress of computers, it has become popular to solve flow problems numerically. By such means it is now possible to follow a kaleidoscopic change of flow.

This field of engineering is referred to as numerical fluid mechanics or computational fluid dynamics. It can be roughly classified into four approaches: the finite difference method, the finite volume method, the finite element method and the boundary element method.

15.1 Finite difference method

15.1.1 Finite difference indication

One of the methods used to discretise the equations of flow for computational solution is the finite difference method.

The fundamental method for indicating a partial differential coefficient in

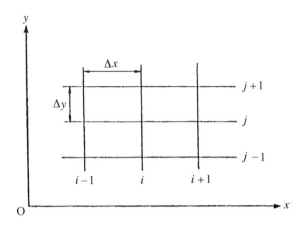

Fig. 15.1 Finite difference method

finite difference form is through the Taylor series expansion of functions of several independent variables. Assume a rectangular mesh, for example. Subscripts (i, j) are to indicate (x, y) respectively as shown in Fig. 15.1. The mesh intervals in the i and j directions are Δx and Δy respectively, while f is a functional symbol. Space points (i, j) mean $(x_i = x_0 + i\Delta x, \, y_i = y_0 + j\Delta y)$.

The forward, backward and central differences of the first-order differential coefficient $\partial f/\partial x$ can be induced in the manner stated below. Provided that function f is continuous, permitting Taylor expansion of f_{i+1} and f_{i-1}, then considering the x direction alone,

$$f_{i+1} = f_i + \left.\frac{\partial f}{\partial x}\right|_i \Delta x + \frac{1}{2}\left.\frac{\partial^2 f}{\partial x^2}\right|_i \Delta x^2 + \frac{1}{6}\left.\frac{\partial^3 f}{\partial x^3}\right|_i \Delta x^3 + \dots \qquad (15.1)$$

$$f_{i-1} = f_i - \left.\frac{\partial f}{\partial x}\right|_i \Delta x + \frac{1}{2}\left.\frac{\partial^2 f}{\partial x^2}\right|_i \Delta x^2 - \frac{1}{6}\left.\frac{\partial^3 f}{\partial x^3}\right|_i \Delta x^3 + \dots \qquad (15.2)$$

Solving eqn (15.1) for $\partial f/\partial x|_i$,

$$\left.\frac{\partial f}{\partial x}\right|_i = \frac{f_{i+1} - f_i}{\Delta x} + O(\Delta x) \qquad (15.3)$$

Here, $O(\Delta x)$ means the combination of terms of order Δx or less. Since this finite difference approximation, omitting $O(\Delta x)$, is approximated by the functional value f_i of x_i and functional value f_{i+1} at x_{i+1} on the side of increasing x, it is called the forward difference. This finite difference indication has a truncation error of the order Δx and it is said to have first-order accuracy. The backward difference is approximated by the functional value f_{i-1} on the side of decreasing x and f_i through a similar process, and

$$\left.\frac{\partial f}{\partial x}\right|_i = \frac{f_i - f_{i-1}}{\Delta x} + O(\Delta x) \qquad (15.4)$$

Furthermore, solving eqns (15.1) and (15.2) for $\partial f/\partial x|_i$, then by subtraction,

$$\frac{\partial f}{\partial x}\bigg|_i = \frac{f_{i+1} - f_{i-1}}{2\Delta x} + O(\Delta x^2) \qquad (15.5)$$

Since this finite difference representation is approximated by functional values f_{i-1} and f_{i+1} on either side of x_i, it is called the central difference. As seen from eqn (15.5), the central difference is said to have second-order accuracy. This method of representation is also applicable to the differential coefficient for y.

Next, the central difference for $\partial^2 f/\partial x^2|_i$ is obtainable by adding eqn (15.1) to eqn (15.2). In other words, it has second-order accuracy:

$$\frac{\partial^2 f}{\partial x^2}\bigg|_i = \frac{f_{i+1} - 2f_i + f_{i-1}}{2\Delta x^2} + O(\Delta x^2) \qquad (15.6)$$

In this way, a partial differential coefficient is expressed in finite difference form as an algebraic equation. By substituting these coefficients a partial differential equation can be converted to an algebraic equation.

15.1.2 Incompressible fluid

Method using stream function and vorticity
To begin with, an explanation is given of the case where the flow pattern is obtained for the two-dimensional steady laminar flow of an incompressible and viscous fluid in a sudden expansion of a pipe as shown in Fig. 15.2. In this case, what governs the flow are the Navier–Stokes equations and the continuity equation.

In the steady case, a vorticity transport equation is derived from the Navier–Stokes equation and is expressed in non-dimensional form. It produces the following equation by putting $\partial\zeta/\partial t = 0$ in eqn (6.18) and additionally substituting the relationship of eqn (12.12), $u = \partial\psi/\partial y$, $v = -\partial\psi/\partial x$:

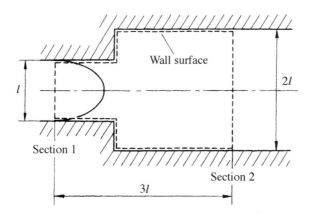

Fig. 15.2 Flow in a sudden expansion

$$\frac{\partial \psi}{\partial y}\frac{\partial \zeta}{\partial x} - \frac{\partial \psi}{\partial x}\frac{\partial \zeta}{\partial y} = \frac{1}{Re}\left(\frac{\partial^2 \zeta}{\partial x^2} + \frac{\partial^2 \zeta}{\partial y^2}\right) \tag{15.7}$$

Also, the vorticity definition equation (4.7) can be expressed in terms of the stream function ψ using the relationship in eqn (12.12):

$$\frac{\partial^2 \psi}{\partial x^2} + \frac{\partial^2 \psi}{\partial y^2} = -\zeta \tag{15.8}$$

The actual numerical computation is made by approximating the above partial differential equations by finite difference equations. In this computation, since the flow appearing in Fig. 15.2 is symmetric about the centre line, only the lower half of the pipe is the computational area and it is covered by a parallel mesh of interval h as shown in Fig. 15.3. Using eqns (15.5) and (15.6) for ψ,

$$\left.\begin{array}{l} \dfrac{\partial \psi}{\partial x} \approx \dfrac{\psi_{i+1,j} - \psi_{i-1,j}}{2h} \\[2mm] \dfrac{\partial \psi}{\partial y} \approx \dfrac{\psi_{i,j+1} - \psi_{i,j-1}}{2h} \\[2mm] \dfrac{\partial^2 \psi}{\partial x^2} \approx \dfrac{\psi_{i-1,j} - 2\psi_{i,j} + \psi_{i+1,j}}{h^2} \\[2mm] \dfrac{\partial^2 \psi}{\partial y^2} \approx \dfrac{\psi_{i,j-1} - 2\psi_{i,j} + \psi_{i,j+1}}{h^2} \end{array}\right\} \tag{15.9}$$

A similar approximate equation to eqn (15.9) is obtained for ζ. Substitute these into eqns (15.7) and (15.8) and rearrange for ζ_{ij} and ψ_{ij} respectively,

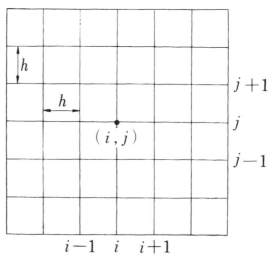

Fig. 15.3 Grid mesh and grid points

$$\zeta_{i,j} = \tfrac{1}{4}(\zeta_{i-1,j} + \zeta_{i,j-1} + \zeta_{i+1,j} + \zeta_{i,j+1})$$

$$+ \frac{Re}{16}[(\psi_{i+1,j} - \psi_{i-1,j})(\zeta_{i,j+1} - \zeta_{i,j-1})$$

$$- (\psi_{i,j+1} - \psi_{i,j-1})(\zeta_{i+1,j} - \zeta_{i-1,j})] \qquad (15.10)$$

$$\psi_{i,j} = \tfrac{1}{4}(\psi_{i-1,j} + \psi_{i,j-1} + \psi_{i+1,j} + \psi_{i,j+1} + h^2\zeta_{i,j}) \qquad (15.11)$$

Equations (15.10) and (15.11) show the relationship between vorticity ζ_{ij} (as well as stream function ψ_{ij}) at mesh points (i, j) in Fig. 15.3 and the vorticities (as well as stream functions) at the surrounding mesh points. If they are described for all mesh points, simultaneous equations are obtained. In general, because such equations have many unknowns and are also non-linear, they are mostly solved by iteration. In other words, substitute into eqns (15.10) and (15.11) the given values of the boundary condition on inlet section 1, the centre line and the wall face for ζ and ψ. Set the initial value for the mesh points inside the area to zero. The values of ζ and ψ will be new values other than zero when their equations are first evaluated. Repeat this procedure using these new values and the value obtained by extrapolating the unknown boundary value on outlet section 2 from the value at the upstream inner mesh point. When satisfactory convergent mesh point values are reached, the computation is finished. Figure 15.4 shows the streamlines and the equivorticity lines in the pipe obtained through this procedure when $Re = 30$.

This iteration method is called the Gauss–Seidel sequential iteration method. Usually, however, to obtain a stable solution in an economical number of iterations, the successive over-relaxation (SOR)[1] method is used.

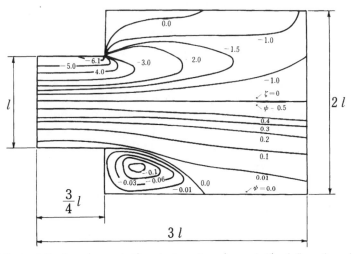

Fig. 15.4 Equivorticity lines (upper half) and streamlines (lower half) of flow through sudden expansion

[1] Forsythe, G. E. and Wasow, W. R., *Finite-Difference Methods for Partial Differential Equations*, (1960), 144, John Wiley, New York.

Furthermore, when the left-hand side of eqn (15.7) is discretised using central differences, a stable convergent solution is hard to obtain for flow at high Reynolds number. In order to overcome this, the upwind difference method[2] is mostly used for this finite difference method.

This method is based upon the idea that most flow information comes from the upstream side. For example, if the central difference is applied to $\partial \psi / \partial y$ of the first term of left side but the upwind difference to $\partial \zeta / \partial x$, then the following equations are obtained.

$$\frac{\partial \psi}{\partial y} = \frac{\psi_{i,j+1} - \psi_{i,j-1}}{2h} \tag{15.12}$$

and

$$\left. \begin{aligned} \frac{\partial \zeta}{\partial x} &= \frac{\zeta_{i,j} - \zeta_{i-1,j}}{h} \quad (\psi_{i,j+1} \geq \psi_{i,j-1}, \text{when } u_i \geq 0) \\ &= \frac{\zeta_{i+1,j} - \zeta_{i,j}}{h} \quad (\psi_{i,j+1} < \psi_{i,j-1}, \text{when } u_i < 0) \end{aligned} \right\} \tag{15.13}$$

Equation (15.13) is still only of first order accuracy and so numerical errors can accumulate, sometimes strongly enough to invalidate the solution.

Method using velocity and pressure

In the preceding section, computation was done by replacing the flow velocity and pressure with the stream function and vorticity to decrease the number of dependent variables. In the case of complex flow or three-dimensional flow, however, it is difficult to establish a stream function on the boundary. In such a case, computation is done by treating the flow velocity and pressure in eqns (6.2) and (6.12) as dependent variables. Typical of such methods is the MAC (Marker And Cell) method,[3] which was developed as a numerical solution for a flow with a free surface, but was later improved to be applicable to a variety of flows. In the early development of the MAC method, markers (which are weightless particles indicating the existence of fluid) were placed in the mesh unit called a cell, as shown in Fig. 15.5, and such particles were followed. One of the examples is shown in Fig. 15.6, where a comparison was made between the photograph when a liquid drop fell onto a thin liquid layer and the computational result by the MAC method.[4,5]

More recently, however, a technique with the variables of flow velocity and pressure separately located (using a staggered mesh) as shown in Fig. 15.7 was adapted from the MAC method. Markers are not needed but are used only for the presentation of results.

[2] Gosman, A. D. *et al.*, *Heat and Mass Transfer in Recirculating Flow*, (1969), 55, Academic Press, New York.
[3] Harlow, F. H. and Welch, J. E., *The Physics of Fluids*, 8, (1965), 2182.
[4] Nakayama, Y. and Nakagome, H., (photograph only).
[5] Nichols, B. D., *Proc. 2nd Int. Conf. on Numerical Methods in Fluid Dynamics*, (1971), 371.

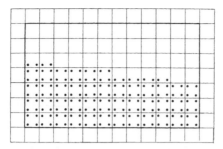

Fig. 15.5 Layout of cell and marker particles used for computing flow on inclined free surface

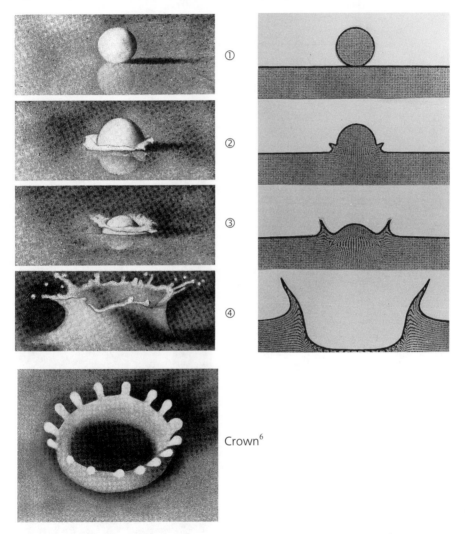

Crown[6]

Fig. 15.6 Liquid drop falling onto thin liquid layer: ① start; ② at 0.0002 s; ③ at 0.0005 s; ④ at 0.0025 s

[6] Fujii, K. and Nakagome, H., *Reading Physical Phenomena* (1978), 102, Kodansha, Tokyo (in Japanese).

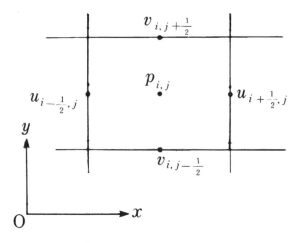

Fig. 15.7 Layout of variables in the MAC method

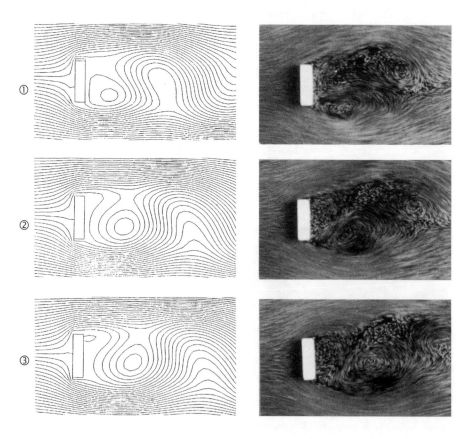

Fig. 15.8 Time-sequenced change of Kármán vortex street: ① start; ② at 0.1 s; ③ at 0.2 s

As an example, in Fig. 15.8 comparison is made between the kaleidoscopic change of Kármán vortices in the flow behind a prism and the computational result.[7]

15.1.3 Compressible fluid

Time-marching method
For a compressible fluid, the equation of a thermodynamic quantity in addition to the equations of continuity and momentum must be evaluated. One-dimensional isentropic flows etc. are solvable analytically. However, the development of a multi-dimensional shock wave, for example, can be solved by numerical methods only. For example, in the MacCormack method,[8] the differential equation is developed from the conservation form[9] for the mass, momentum and energy, neglecting the viscosity.

Figure 15.9 is the equi-Mach-number diagram of a rocket head flying at supersonic velocity calculated by using this method.[10]

One of the methods used to solve the compressible Navier–Stokes equation taking the viscosity into account is the IAF (Implicit Approximate

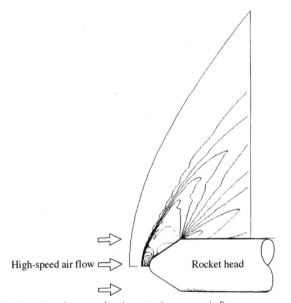

High-speed air flow ⟹   Rocket head

Fig. 15.9 Equi-Mach number diagram of rocket nose in supersonic flow

[7] Nakayama, Y., Aoki, K. and Oki, M., *Proc. 3rd Asian Symp. on Visualization*, (1994), 453.
[8] MacCormack, R. W., *AIAA Paper*, 69-354, (1969).
[9] The conservation form of a one-dimensional inviscid compressible fluid is

$$\frac{\partial f}{\partial t} + \frac{\partial g}{\partial x} = 0 \quad f = \left\{ \begin{array}{c} \rho \\ \rho u \\ e \end{array} \right\} \quad g = \left\{ \begin{array}{c} \rho u \\ p + \rho u^2 \\ u(e + p) \end{array} \right\}$$

[10] Hirose, N. *et al.*, National Aerospace Lab., Japan.

Factorisation) method which is sometimes called the Beam–Warming method.[11] In Fig. 15.10 it is applied to a transonic turbine cascade. The solution is produced by using this method only for the region near the turbine cascade, while using the finite element method for the other region. Results matching the test result well are obtained.[12] As an example of a three-dimensional case, Plate 5[13] shows the result obtained by solving the compressible Navier–Stokes equation for the density distribution of the flow on the rotating fan blades and spinner of a supersonic turbofan engine by the IAF method.

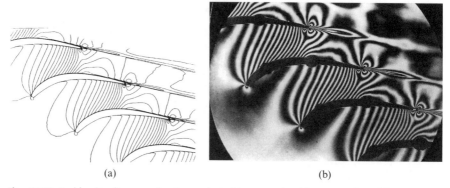

(a)   (b)

Fig. 15.10 Equidensity diagram of a transonic turbine cascade: (a) computation; (b) experiment (photograph of Mach–Zehnder interference fringe)

Method of characteristics

Figure 15.11 is a test rig for water hammer, which is capable of measuring the pressure response waveform by the pressure transducer set just upstream of the switching valve. When the switching valve is suddenly closed, pressure p increases and propagates along the pipe as a pressure wave. To obtain its numerical solution, the wave phenomenon is expressed by a hyperbolic equation, and the so-called method of characteristics[14] is used.

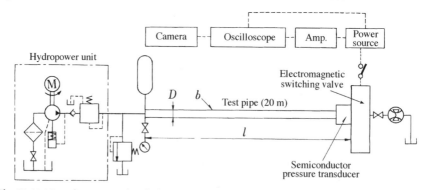

Fig. 15.11 Water hammer testing device

[11] Beam, R. M. and Warming, R. F., *AIAA Journal*, 16 (1978), 393.
[12] Nakahashi, K. *et al.*, *Transactions of the JSME*, 54, (1988), 506.
[13] Nozaki, O. *et al.*, *Proc. Int. Symp. on Air Breathing Engines*, (1993).
[14] Steerer, V. L., *Fluid Mechanics*, (1975), 6th edition, 654, McGraw-Hill, New York.

Now, putting f as the friction coefficient of the pipe and a as the propagation velocity of the pressure wave, linearly combine the continuity equation, which is the one-dimensionalised equations (6.1) and (6.12), with λ times the momentum equation, to get

$$\frac{\lambda}{\rho a^2}\left[\frac{\partial p}{\partial t}+\left(v+\frac{a^2}{\lambda}\right)\frac{\partial p}{\partial x}\right]+\left[\frac{\partial v}{\partial t}+(v+\lambda)\frac{\partial v}{\partial x}\right]+\frac{f}{2D}v|v|=0 \qquad (15.14)$$

Here, assume that

$$v+\frac{a^2}{\lambda}=\frac{dx}{dt} \qquad v+\lambda=\frac{dx}{dt} \qquad (\lambda=\pm a) \qquad (15.15)$$

and partial differential equation (15.14) is converted to an ordinary differential equation. Furthermore, discretise it, and, as shown in Fig. 15.12, v and p of point P after time interval Δt are obtained as the intersection of the curves C^+ ($\lambda=a$) and C^- ($\lambda=-a$) which are expressed by eqn (15.15) from the initial values of velocity v and pressure p at A and C.

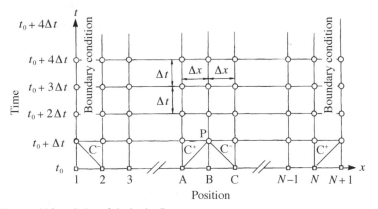

Fig. 15.12 x–t grid for solution of single pipe line

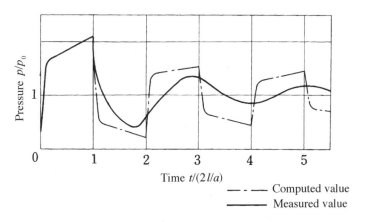

Fig. 15.13 Pressure response wave in water hammer action

Figure 15.13 shows the comparison between the pressure waves thus calculated and the actually measured values.[15] The difference between them arises from the fact that the frequency dependent pipe friction is not taken into account in eqn (15.14).

15.1.4 Turbulence

Turbulence model

As already stated in Section 6.4, making some assumption or simplification for computing the Reynolds stress τ_t, expressed by eqn (6.39), is called the modelling of turbulence. It is mainly classified by the number of transport equations for the turbulence quantity used for computation. The equation for which τ_t is given by eqn (6.40) or (6.43) is called a zero-equation model. The equation for which the kinetic energy k of turbulence is determined from the transport equation, while the length scale l of turbulence is given by an algebraic expression, is called a one-equation model. And the method by which both k and l are determined from the transport equation is called a two-equation model. The k–ε model, using the turbulence energy dispersion ε instead of l, is typical of the two-equation model. As an example, Fig. 15.14 shows the mesh diagram used to compute the flow in a fluidic device and also the computational results of streamline, turbulence energy and turbulence dispersion.[16]

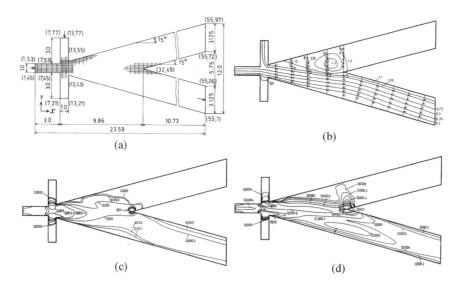

Fig. 15.14 Flow in a fluidic device: (a) mesh diagram; (b) streamline; (c) turbulent energy; (d) turbulent dispersion. $Re = 10^4$, $Q_c/Q_s = 0.2$ (Q_c: control flow rate; Q_s, supply flow rate)

[15] Izawa, K., MS thesis, Faculty of Engineering, Tokai University, (1976).
[16] Ogino, H. and Nakayama, Y., *Bulletin of the JSME*, 29 (1986), 1515.

LES (Large Eddy Simulation)

In computations based on the time-averaged Navier–Stokes equation using turbulence models, time is averaged and the change in turbulence is treated as being smooth. However, a method by which computation can follow the change in irregularly changing turbulence for clarifying physical phenomena etc. is LES.

LES is a method where the computation is conducted by modelling only vortices small enough to stay inside the mesh in terms of local mean (mesh mean model), while large vortices are not modelled but computed as they are. Figure 15.15(a) shows a solution for the flow between parallel walls.[17] Comparing this with Fig. 15.15(b), a visualised photograph of bursts by the

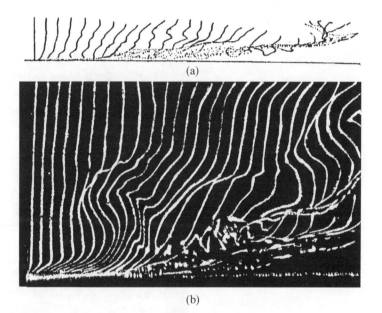

(a)

(b)

Fig. 15.15 Time lines near the wall of a flow between parallel walls: (a) computed; (b) experimental

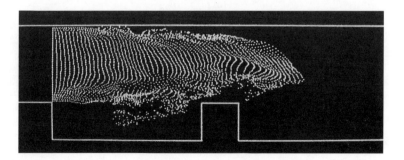

Fig. 15.16 Turbulent flow over step (large eddy simulation). Reynolds number based on a channel width, $Re = 1.1 \times 10^4$.

---

[17] Moin, P. and Kim, J., *Journal of Fluid Mechanics*, 118, (1982), 341.

hydrogen bubble method,[18] it is clear that they coincide well with each other. In Fig. 15.16, the turbulent flow over a step is computed and its time lines are shown graphically.[19] Plate 2 shows the computational result for turbulent flow around a rectangular column.[20]

Direct simulation

If the Navier–Stokes equation and continuity equation are computed directly as they are, then turbulence can be computed without using a model. This is called the direct simulation of turbulence. Even with the number of mesh points available in the latest large computer, only the larger turbulent vortices can be found. Nevertheless, interesting results on the large structure of turbulence have been obtained.[21]

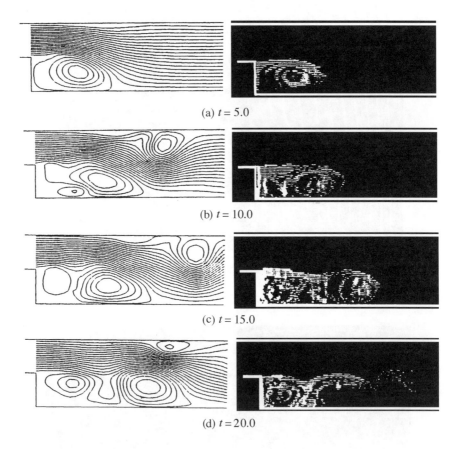

(a) $t = 5.0$

(b) $t = 10.0$

(c) $t = 15.0$

(d) $t = 20.0$

Fig. 15.17 Flow behind a step

[18] Kim, H. T. *et al.*, *Journal of Fluid Mechanics*, 50, (1971), 113.
[19] Kobayashi, T. *et al.*, Report IIS, University of Tokyo, 33 (1987), 25.
[20] Kobayashi, T., *Atlas of Visualization III*, Plate 10, (1997), CRC Press, Boca Raton, FL.
[21] Kuwahara, K., *Simulation of Turbulence, Journal of Japan Physics Academy*, 40, (1985), 877.

These methods simulate the movement of a large vortex by making the accuracy of the upwind difference scheme, shown in Fig. 15.13, of higher order and also by making the numerical viscosity[22] smaller. As one such example, the computed and visualised flows behind a step are shown in Fig. 15.17.[23] It can be seen that the movement of the vortex behind the step with the passage of time is well simulated.

15.2 Finite volume method

The finite volume method is a technique which discretises in a small region (the control volume shown in Fig. 15.18) the integration equation of the continuity equation and the Navier–Stokes equation written in conservative form.[24] The boundary volumes are then obtained using the neighbouring grid points.[25]

In the examples which appeared in the preceding sections, the grid was a regular structured grid in a line. Of late, however, the boundary-fitted grid following an irregular boundary or an unstructured grid has also been used. In the finite volume method, these new grids are easier to apply. As examples, the application of these techniques to an unstructured grid of triangles, the

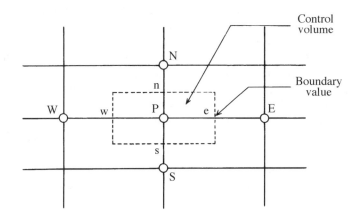

Fig. 15.18 Control volume

---

[22] This means the artificial propagation term produced by the finishing error of the upwind differential.
[23] Oki, M. *et al.*, *JSME International Journal*, 36-4, B (1993), 577.
[24] For example, the Navier–Stokes equation written in preservative form is obtained by expressing $u\partial u/\partial x$, $v\partial u/\partial y$, etc., the inertia term of eqn (16.12), in the form of $\partial(u \cdot u)/\partial x$, $\partial(u \cdot v)/\partial y$.
[25] Patankar, S. V., *Numerical Heat Transfer and Fluid Flow*, (1980), Hemisphere, New York.

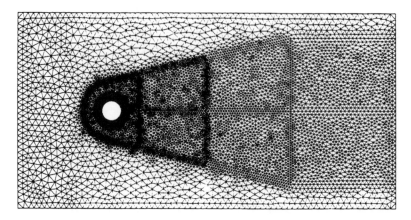

Fig. 15.19 Unstructured grid[26]

flow around a column, the mesh and the computed pressure distribution and velocity vector diagram are shown in Fig. 15.19 and Plate 1.

15.3 Finite element method

15.3.1 Division of elements

The finite difference method is a mathematical method by which the differential calculus appearing in the governing equation is directly approximated by finite difference equations. In the finite element method, however, by using physical approximations to discretise the differential equations, simultaneous algebraic equations are developed for the whole elements. Thus an approximate solution of the differential equations satisfying the boundary conditions is obtained. The flow zone was divided into a right-angled mesh as a rule in the finite difference method. In the finite element method, however, by dividing the area into proper-sized triangular or quadrangular elements as shown in Fig. 15.20, any complex-shaped area can be treated. The corners of the triangles or quadrangles are called nodal points, at which such variables as x, y, u, v and p are defined.

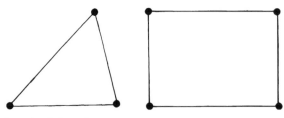

Fig. 15.20 Two-dimensional elements

[26] Oki, M. *et al.*, *Trans. JSME*, 65–631, B (1999), 870.

15.3.2 Method of weighted residuals

For discretisation by the finite element method, the variational principle or the method of weighted residuals is used. The variational principle is also called the minimum energy principle, which uses the principle that the potential energy is a minimum in the state of equilibrium. As this method has limited application, the method of weighted residuals is widely used.

Consider the potential flow around a cylinder placed between flat plates as shown in Fig. 15.21.

$$
\left.
\begin{aligned}
\text{in region S containing fluid} \quad & \frac{\partial^2 \psi}{\partial x^2} + \frac{\partial^2 \psi}{\partial y^2} = 0 \\[2mm]
\text{At inlet and on wall surface } S_1 \quad & \psi = \overline{\psi} \\[2mm]
\text{At outlet } S_2 \text{ which is free boundary} \quad & \frac{\partial \psi}{\partial n} = \frac{\overline{\partial \psi}}{\partial n}
\end{aligned}
\right\}
\qquad (15.16)
$$

where the bars above the letters indicate that the applicable values are those on the boundary.

Next, in order to obtain the stream function ψ, multiply by a given function which is $\psi^* = 0$ on boundary S_1 (and can be any value in other areas by eqn (15.16)). Then integrate for the whole region. The following equation is obtained:

$$
\int_S \left(\frac{\partial^2 \psi}{\partial x^2} + \frac{\partial^2 \psi}{\partial y^2} \right) \psi^* \, dA + \int_{S_2} \left(\frac{\overline{\partial \psi}}{\partial n} - \frac{\partial \psi}{\partial n} \right) \psi^* \, dS = 0
\qquad (15.17)
$$

Here, function ψ^* is called the weighting function. In eqn (15.17), assume function ψ^* and its derivative $\partial \psi / \partial n$ are approximate values. The first term on the left expresses the quantity obtained by multiplying the error of the differential equation in the area (here, called the residual) by a given function and integrating for the whole area. Likewise, the second term expresses the quantity obtained by applying a similar process to the residual on boundary S_2. This is called a weighted residual expression. When the right solution is

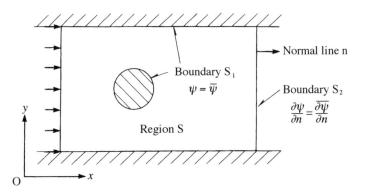

Fig. 15.21 Flow around cylinder

obtained, this equation applies strictly to the given function ψ^*. The approximate solution which distributes the error to satisfy the function $\psi^* = 0$ is called the method of weighted residuals.

15.3.3 Interpolating function

In the finite element method, improvement is made by applying an algebraic equation derived using the values at nodal points to approximate the unknowns in each element. This equation is called an interpolating function. Where a weighting function of the same type is chosen it is called the Galerkin method.

It is not easy to obtain an approximate function effective all over sections [a, b] for the one-dimensional function $\psi = \psi(x)$ shown in Fig. 15.22. Nevertheless, the section [a, b] can be divided into large and small linear elements. For example, divide the subsection where the function changes abruptly into (1, 2), and divide the subsection of the gentler change into (3, 4). Then for each of them ψ can be expressed by a one-dimensional (linear) function.

In the two-dimensional case, as shown in Fig. 15.23, by using triangular elements their size can be determined to the extent that the functions are expressible by a one-dimensional function of coordinates according to how abruptly or gently the functional change is expected. In other words,

$$\psi = \alpha_1 + \alpha_2 x + \alpha_3 y \tag{15.18}$$

Assume the function values at the corners of triangle 1, 2 and 3 to be ψ_1, ψ_2 and ψ_3 respectively, then

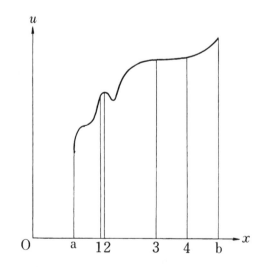

Fig. 15.22 One-dimensional function

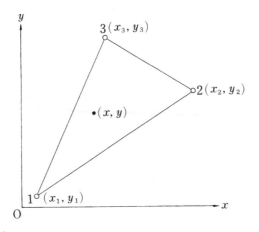

Fig. 15.23 Triangular element

$$\begin{Bmatrix} \psi_1 \\ \psi_2 \\ \psi_3 \end{Bmatrix} = \begin{bmatrix} 1 & x_1 & y_1 \\ 1 & x_2 & y_2 \\ 1 & x_3 & y_3 \end{bmatrix} \begin{Bmatrix} \alpha_1 \\ \alpha_2 \\ \alpha_3 \end{Bmatrix} \qquad (15.19)$$

From the above,

$$\begin{Bmatrix} \alpha_1 \\ \alpha_2 \\ \alpha_3 \end{Bmatrix} = \begin{bmatrix} 1 & x_1 & y_1 \\ 1 & x_2 & y_2 \\ 1 & x_3 & y_3 \end{bmatrix} \begin{Bmatrix} \psi_1 \\ \psi_2 \\ \psi_3 \end{Bmatrix} \qquad (15.20)$$

Substitute eqn (15.20) into (15.18),

$$\psi = \phi_1\psi_1 + \phi_2\psi_2 + \phi_3\psi_3 = \sum_{i=1}^{3} \phi_i\psi_i \qquad (15.21)$$

In other words, ψ is the interpolating function expressed as the linear combination of nodal point values ψ_i. Hence, in the following form,

$$\phi_i = a_i + b_ix + c_iy \quad (i = 1, 2, 3) \qquad (15.22)$$

it is called the shape function, and a_i, b_i and c_i are determined by the coordinates of the nodal points.

15.3.4 Equation-overlapping elements

Approximate the unknown function ψ and weighting function ψ^* respectively in eqn (15.17) by interpolating the functional equation (15.21) using the nodal point values in the element and the same equation with ψ changed to ψ^*. Substituting these functions into the weighted residual equation, which is the deformed equation (15.17), gives the quantitative relation for each element. By overlapping them, a simulated linear equation covering the whole analytical area is developed. By solving these equations, it is possible to obtain the values at each nodal point and thus to draw the streamline of ψ = constant.

15.3.5 Applicable cases

To compute the flow shown in Fig. 15.21, as this is the symmetrical flow, the upper half only of the flow is divided into large and small triangular elements as shown in Fig. 15.24. For the finite element method, it is enough, unlike the finite difference method, just to divide the flow section finely around the cylinder where the velocity changes abruptly.

The computed streamline and velocity vector are shown in Fig. 15.25.[27]

With the finite element method also, as for the finite difference method, analysis of viscous and compressible fluids is possible. More recently, computation using a turbulence model has been carried out. As examples for a viscous fluid, the computational result for laminar flow around a pipe nest is shown in Fig. 15.26,[28] while that for the turbulence velocity distribution of the flow in a clean room using the k–ε model is shown in Plate 3.[29]

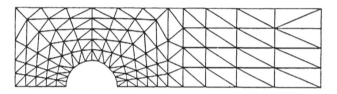

Fig. 15.24 Mesh diagram of flow around cylinder (180 elements and 115 nodes)

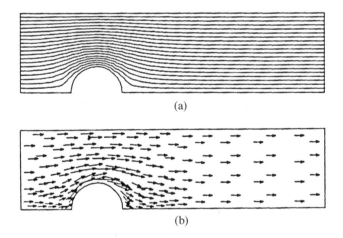

(a)

(b)

Fig. 15.25 Flow around cylinder: (a) streamline; (b) velocity vector

[27] Hayashi, K. *et al.*, *Flow Analysis by Personal Computer*, (1986), 73, Asakura-Shoten, Tokyo.
[28] Nakazawa, J., *Journal of JSME*, 87 (1984), 316.
[29] Ikegawa. M. *et al.*, *Proc. Int. Symp. on Supercomputers for Mechanical Engineering, JSME*, (1988), 57.

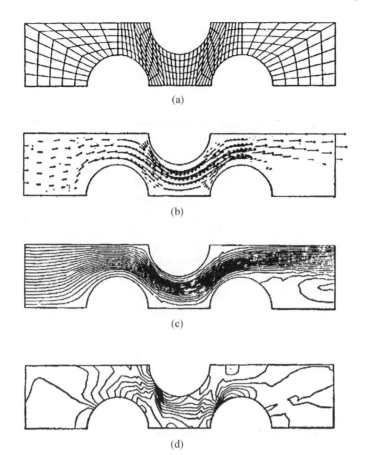

Fig. 15.26 Flow around tube bank: (a) divided element; (b) velocity vector; (c) streamline; (d) pressure (Re = 100)

15.4 Boundary element method

Instead of solving the difference equation which governs fluid movement under the given boundary conditions, the boundary element method uses an integral equation which must satisfy values on the boundary. To derive the integral equation, one can use the method using Green's formula and also the method of weighted residuals. Green's formula method has long been used for analysing potential flow, and more recently has been systematised as the 'panel method', used for analysing external flows around aircraft, automobiles, etc.

Brebbia derived an equation by the more general method of weighted residuals with wider applicability, and named it the boundary element method.[30] It is often compared with the finite element method, and has been used in many fields of application.

[30] Brebbia, C. A., *The Boundary Element Method for Engineers*, (1978), Pentech Press, London.

In this method, the weighting function in the method of weighted residuals described in Section 15.3.2 is selected so as to satisfy the Laplace equation (15.16) within area S, and converted to an integral equation on boundary S_2 surrounding the area as shown by the following equation:

$$\int_{S_2} \psi^* \frac{\partial \psi}{\partial n} dS - \int_{S_2} \psi \frac{\partial \psi^*}{\partial n} dS = 0 \tag{15.23}$$

Next, the boundary is divided into a number of line-segment elements. For example, in the case of the flow shown in Fig. 15.24, the mesh division is as shown in Fig. 15.27. Then, the value at a given point in the element is expressed in terms of the value of the nodal point by the interpolating equation (15.21) in the finite element method. The simultaneous linear equation for the value at the nodal points can then be solved.

The computational result for the case of Fig. 15.27 is shown in Fig. 15.28.[30] Here, $\partial \psi / \partial n$ expresses the flow velocity along the boundary.

Since the boundary element method only requires division of the boundary of the region into the elements, it is popular for cases where the velocity or the pressure distribution on a body surface needs to be obtained.

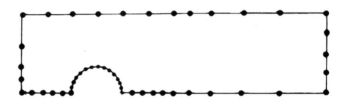

Fig. 15.27 Mesh diagram by boundary element method of flow around cylinder

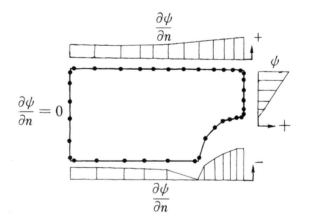

Fig. 15.28 Solution by boundary element method

[30] Brebbia, C. A., *The Boundary Element Method for Engineers*, (1978), Pentech Press, London.

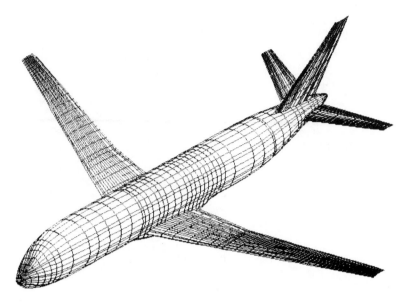

Fig. 15.29 Mesh diagram for computing flow around full model of transonic plane

Figure 15.29 is the mesh diagram for the case of flow around a full model of a transonic plane using the panel method. The computational result of the pressure distribution obtained is shown in Plate 6(a), which coincides very well with the result of the wind tunnel experiment as shown in Plate 6(b).[31]

Finally, a new kind of finite volume method has been proposed. This

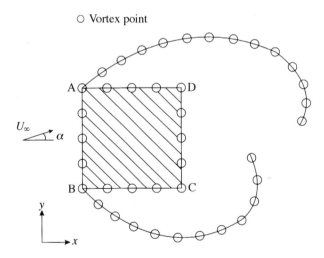

Fig. 15.30 Modelling by discrete vortex element

[31] Kaiden, T. *et al.*, *Proc. 6th NAL Symp. on Aircraft Computational Aerodynamics*, (1988), 141.

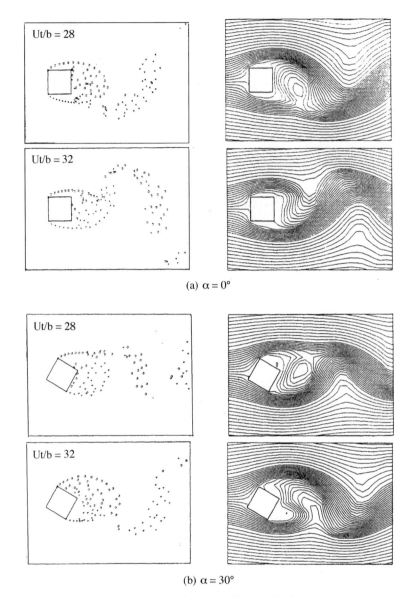

Ut/b = 28

Ut/b = 32

(a) α = 0°

Ut/b = 28

Ut/b = 32

(b) α = 30°

Fig. 15.31 Flow pattern around the rectangular column illustrated in Fig. 15.30

technique replaces the successive distribution of vorticity produced in a flow field containing varied viscosity and density with discrete vortex elements. Each vortex motion is followed by the Lagrange method and thus analyses the unsteady flow field. This technique is called the discrete vortex method. As an example, the computational results for an unsteady flow around a square column in a uniform flow are shown in Fig. 15.30.[32]

---

[32] Inamoto, T. *et al.*, *Finite Element Flow Analysis*, University of Tokyo Press, (1982), 931.

In Fig. 15.31(a) and (b) the left and right sides show respectively the flow pattern of vortex points and streamlines. In any of these cases, the positive vortex (clockwise rotation) develops from point A and the negative vortex (counterclockwise rotation) from points B and C. These vortices develop behind the rectangular column and the Kármán vortex street is formulated in the wake.

16

Flow visualisation

The flow of air cannot be seen by the naked eye. The flow of water can be seen but not its streamlines or velocity distribution. The consolidated science which analyses the behaviour of fluid invisible to the eye like this as image information is called 'flow visualisation', and it is extremely useful for clarifying fluid phenomena. The saying 'seeing is believing' most aptly expresses the importance of flow visualisation. Analytical studies clarifying hitherto unclear flows and the developmental studies of flows in and around machinery have been much assisted by this science.

About a century ago, Reynolds made the great discovery of the law of similarity by visualisation. Thereafter, Prandtl's concept of the boundary layer and his ideas for its control, Kármán's clarification of his vortex street, Kline's discovery of the bursting phenomenon allied to developing the mechanism of turbulence, and other major discoveries concerned with fluid phenomena were mostly achieved by flow visualisation. Furthermore, in the clarification of turbulent structure, the establishment of mathematical models of turbulence, etc., which currently still pose big problems, flow visualisation is furnishing extremely important information

In recent years, with the progress of computers, its use has been enhanced by image processing. Also, computer-aided flow visualisation (CAFV), the image presentation of numerical computations and measured results, is making great advances.

16.1 Classification of techniques

The visualisation techniques are classified as shown in Table 16.1 and divided roughly into experimental methods and computer-aided visualisation methods.

16.2 Experimental visualisation methods

16.2.1 Wall-tracing method

The oil-film method, typical of this technique, has long been used, so the technique is well established. There are many applications, and it is used for

Table 16.1 Classification of visualisation techniques

Visualisation technique	Air flow	Water flow	Explanation
Experimental visualisation method			
1. Wall surface tracing method			
Oil-film method, oil-dots method	○	●	By attaching oil film or oil dots to the body surface, from the stream pattern generated, the state including the direction of the flow can be visualised
Mass transfer method	○	●	By utilising the dissolution, evaporation or sublimation into the fluid of a film of a substance attached to the body, the flow state on the body surface can be visualised
Electrolytic corrosion method	○	●	By utilising the corrosion due to electrolysis, the flow state on the body surface can be visualised
Temperature-sensitive-film method	○	●	The surface temperature is visualised according to the colour distribution of a liquid crystal or such attached to the body
Pressure-sensitive-paint method	○	●	By utilising the luminescence of a substance applied to the body surface, the pressure distribution on the surface can be visualised
Pressure-sensitive-paper method	○	●	By utilising the colour density of the pressure-sensitive paper, the pressure distribution on the body surface can be visualised
2. Tuft methods			
Various tuft methods	○	●	Method by which the flow direction is visualised from the flight behaviour of numerous short pieces of thread (tufts). By the surface tuft method the flow near the surface is visualised, while by the depth tuft method the flow at a given point just off the surface, and by the tuft grid method the flow on a given section, and by the tuft stick method the flow at a given point is visualised, respectively
Luminescent mini-tuft method	○	●	Method by which, hardly having any effect on the flow, a single filament of nylon soaked in luminescent dye beforehand is photographed under highly luminous ultraviolet rays

Table 16.1 *Continued*

Visualisation technique	Air flow	Water flow	Explanation
3. Injected tracer method			
Injection streak line method[a]	○	●	Continuously inject tracers, capture the picture at a certain instant, and thus visualise the stream and streak line
Injection path line method[b]	○	●	Intermittently inject tracers for some duration. Visualise the path line
Suspension method[c]	○	●	Evenly suspend liquid or solid particles in the fluid in advance. Thus visualise the stream- and path lines
Surface floating tracer method[d]		●	Let the tracer float on the liquid surface and thus visualise the stream- and streak lines on the liquid surface
Time line method[e]	○	●	Inject the tracer vertically into the flow and thus visualise the time line
4. Chemical reaction tracer method			
Non-electrolytic reaction method	○	●	By utilising the chemical reaction of a fluid with another specified substance, the flow behaviour on the solid surface or the boundary between two fluids can be visualised
Electrolytic colouring method	○	●	By utilising an electrolytically coloured substance as the tracer, the stream line/streak line can be visualised
5. Electric controlled tracer method			
Hydrogen bubble method	○	●	Utilises as the tracer hydrogen bubbles developed through electrolysing with a fine metal wire as the negative pole. Visualise the stream, streak, path and time lines
Spark tracing method	○		Visualise the time line by means of groups of discharge sparks obtained one after another by the high-voltage pulse
Smoke wire method	○		Instantaneously heat an oiled fine metal wire to produce white smoke. Visualise the streak and time lines using the white smoke as the tracer
6. Optical method			
Shadowgraph method	○	●	Let a light emitted from a single point or parallel rays go through a flow region; the flow is visualised by means of the dark and grey shadow thus developed according to the changes in light density

Schlieren photograph method ○ ● Parallel rays are made to deflect through a flow field with a difference in density. The deflected rays are cut with a knife edge, and the density gradient is visualised according to the difference in brightness thus developed

Mach–Zehnder interferometer method ○ ● Parallel rays are divided into two, one of which is made to go through a flow with a difference in density. Quantitative judgement of density and pressure is made from the interference fringe developed by combining the two

Laser holographic interferometer method ○ ● A laser light is separated into two beams. Interference, the fringe pattern, from an object and another beam (reference beam) between the scattered beam is recorded on the hologram film. Illuminating this film by the reference beam, the object can be reconstructed.

Laser light sheet method ○ ● Laser rays are made to strike a cylindrical lens or a revolving or vibrating mirror to make a sheet-like ray, and the three-dimensional flow is visualised as a two-dimensional flow by light scattered from tracer particles

Speckle method ○ ● The flow velocity distribution is obtained by optically processing the speckle pattern obtained by instantaneously photoshooting at short intervals a fluid with suspended tracer particles

Computer-aided visualisation method
7. Visualised image analysing method
PIV (Particle Imaging Velocimetry)

PTV (Particle Tracking Velocimetry) is where the flow velocity distribution is obtained by pursuing every now and then the tracer particles distributed in a fluid relatively thinly in terms of particle density. Correlation method is where the flow velocity distribution is obtained according to the similarity of distribution patterns developing at short intervals of the tracer particles distributed relatively densely in a fluid in terms of particle density

LSV (Laser Speckle Velocimetry) is where the flow velocity distribution is obtained by optically processing the speckle pattern obtained through instantaneous exposure at short intervals of tracer particles suspended in a fluid

HPIV (Holographic PIV) is where three-dimensional velocity information is obtained by recording locational information on holograms and reconstructing it

Table 16.1 *Continued*

Visualisation technique	Explanation
Computer tomography	Such integral value data as the density and temperature from whole cylindrical directions of a given section are collected, and the density and temperature distribution on the section is obtained through computation
Remote sensing	The type and conditions of a body are clarified by capturing the original electromagnetic waves emitted from the body by an aircraft or satellite
Thermographical method	By catching the infrared rays radiated from the liquid surface, the surface temperature can be measured
8. Numerical data visualisation methods	
Contour manifestation method	Physically equal values are connected by a contour
Area colouring manifestation method	Manifestation is made by painting the areas in colours respectively corresponding to their levels of physical quantity
Isosurface manifestation method	The values of physically equal value are three-dimensionally manifested in a surface
Volume rendering method	The levels of equi-value area manifestation are manifested by changing their degree of transparency
Vector manifestation method	The size and direction of the flow velocity vector etc. are manifested in arrow marks
Animation method	Still images on a display are developed into as-if-moving images by continuous shooting
9. Measured data visualisation method	Methods are practised such as those which utilise flow velocimeters and pressure gauges where the velocity distribution, pressure distribution, etc., obtained are to be manifested in an image by processing the data simultaneously, and those which utilise the acoustic intensity method so that the size and direction at every point in the observation arena are manifested in terms of vectors

For injection tracer methods, the names of typical tracers are as follows:
[a] smoke (○), colouring matter (●); [b] soap bubble (○), air bubble (●), oil drop (●), luminescent particle (●); [c] metaldehyde (○), air bubble (○), cavitation (●), liquid tracer (●), aluminium powder (●), polystyrene particle (●); [d] aluminium powder (●), sawdust (●), foaming polystyrene (●); [e] smoke (○), colouring matter (●).

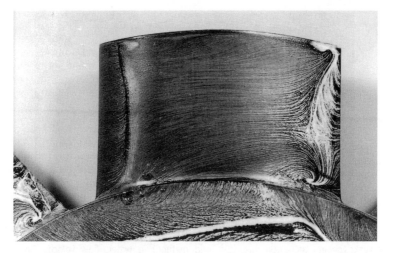

Fig.16.1 Limiting streamlines of Wells turbine for wave power generator (revolving direction is counterclockwise) in water, flow velocity 3.2 m/s, angle of attack 11°

both water and air flow. The flows in the neighbourhood of a body surface, of a wall face inside fluid machinery, etc., have been observed. Figure 16.1 shows the oil-film pattern on the blade surface of a Wells turbine for a wave power generator.[1] From this pattern the nature of the internal flow can be surmised.

16.2.2 Tuft method

Although this is an unsophisticated method widely used for fluid experiments for some time, it has recently become easier to use and more informative as detailed experiments and analyses have been made of the static and dynamic tuft characteristics. It is utilised for visualising flows near and around the surfaces of aircraft, hulls and automobiles as well as those behind them, the internal flows of pumps and blowers, and ventilation flows in rooms. Figure 16.2 shows an example of the visualised flow behind an automobile,[2] while Fig. 16.3 shows that around a superexpress train.[3] Figure 16.4 shows an example of the utilisation of extremely fine fluorescent mini-tufts which hardly disturb the flow.[4]

16.2.3 Injection tracer method

For water flow, the colour streak method has widely been used for a long time. In the suspension method, aluminium powder or polystyrene particles

[1] Tagori, T, *et al.*, *Flow Visualization*, 4, Suppl. (1984), 51.
[2] Tagori, T. *et al.*, *Proc. Flow Visualization Symp.*, (1980), 13.
[3] Japan National Railways.
[4] Saga, T. and Kobayashi, T., *Flow Visualization*, 5, Suppl. (1985), 87.

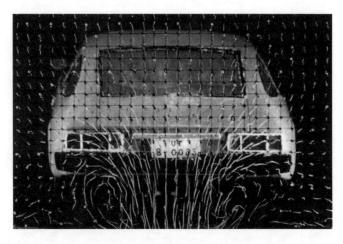

Fig.16.2 Wake behind an automobile (tuft grid method) in water, flow velocity 1 m/s, length 530 mm (scale 1 : 8), $Re = 5 \times 10^5$

Fig.16.3 Flow around a superexpress train (surface tuft method)

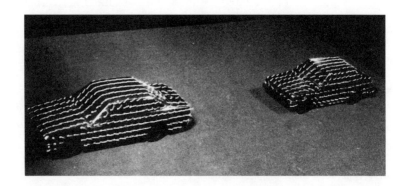

Fig.16.4 Flow around an automobile (fluorescent mini-tuft method)

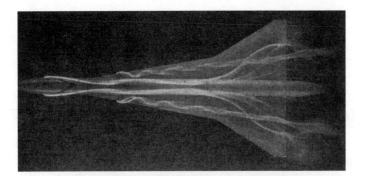

(a)

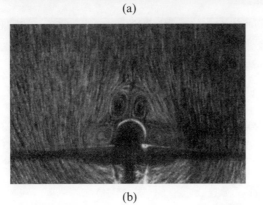

(b)

Fig.16.5 Flow around a double delta wing aircraft in water, angle of attack 15°: (a) colour streak line method; (b) suspension method (air bubble method)

are used, while in the surface floating tracer method, sawdust and aluminium power are used. The smoke method is used for air flows.

There are many examples for visualising the flow around or behind wings, hulls, automobiles, buildings and bridge piers, as well as for the internal flow of pipe lines, blood vessels and pumps.

Figure 16.5 is a photograph where the flow around a double delta wing

Fig.16.6 Flow around an automobile (smoke method)

aircraft is visualised by a water flow.[5] It can be seen how the various vortices develop. These vortices act to increase the lift necessary for a high-speed aircraft to undertake low-speed flight.

Plate 7[6] and Fig. 16.6[7] visualise the flow around an automobile by the smoke method. The flow pattern is clearly seen.

Figure 16.7 shows observation, by the floating sawdust method, of the flow in a bent divergent pipe.[8]

Figure 16.8 visualises a Kármán vortex street using as the tracer the white condensation produced when water is electrolysed with the cylinder as the positive pole.[9]

Fig.16.7 Flow in a bent divergent pipe (floating sawdust method) in water, flow velocity 0.4 m/s, $Re = 2.8 \times 10^4$

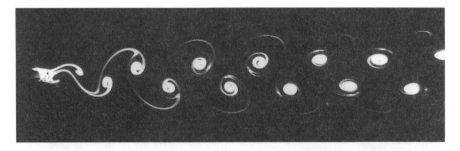

Fig.16.8 Kármán vortex street behind a cylinder (electrolytic precipitation method) in water, flow velocity 10 mm/s, diameter of cylinder 10 mm, $Re = 105$

[5] Werle, H., *Proc. ISFV, Tokyo* (1977), 39.
[6] Flow Visualization Society, Tokyo, *Flow Visualization Handbook*, (1997), 103.
[7] Hucho, W. H. and Janssen, L. J., *Proc. ISFV, Tokyo* (1997), 103.
[8] Akashi, K. *et al.*, *Symp. on Flow Visualization (1st)*, (1973), 100.
[9] Taneda, S., *Fluid Mechanics Learned from Pictures*, Asakura Shoten, (1988), 92.

16.2.4 Chemical reaction tracer method

There are various techniques using chemically reactive substances. Since they have negligible change in density due to chemical reaction, the settling velocity of the tracer is small and thus many of them are suitable for visualising low-velocity flow.

The method has been used for visualising the flow around and behind a flat board, wing and hull, the flow inside a pump and boiler, and natural/thermal convection.

Figure 16.9 is an observation of flow using the streaks developed by injecting saturated liquid ammonium sulphide through a fine tube onto a mixture of white lead and a quick-drying oil which has been applied to the surface of a model yacht.[10]

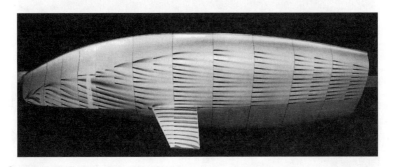

Fig.16.9 Flow on a model yacht surface (surface film colouring method) in water, flow velocity 1.0 m/s, length of model 1.5 m, $Re = 1.34 \times 10^6$, white lead and ammonium sulphide used

16.2.5 Electrically controlled tracer method

Included in this method are three categories: the hydrogen bubble method, spark tracing method and smoke wire method. Any one of them is capable of providing quantitative measurement.

By these methods the flow around and the vortex behind a cylinder, flat board, sphere, wing, aircraft and hull, the flow in a cylinder, the flow around a valve, and the flow in a blower/compressor have been observed.

Plate 8 is a picture visualising the flow around a cylinder by the hydrogen bubble method,[11] while Plate 9 shows the flow around a sphere by the spark tracing method.[12] Figure 16.10 shows the flow around a wing by the same method,[13] and Fig. 16.11 shows the flow around an automobile by the smoke wire method.[14]

[10] Matsui, S., Nishinihon Ryutaigiken Co., Nagasaki, Japan.
[11] Endo, H. et al., Symp. of Flow Visualization (2nd), (1947), 135.
[12] Nakayama, Y., Flow Visualization, 8 (1988), 14.
[13] Nakayama, Y. et al., Symp. on Flow Visualization (4th), (1976), 105.
[14] Nakayama, Y., Faculty of Engineering, Tokai University.

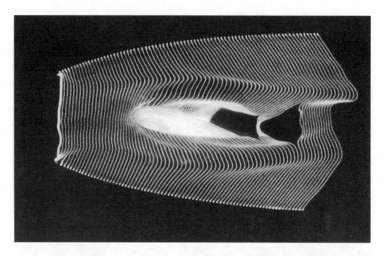

Fig.16.10 Flow around a wing (spark tracing method) in air, flow velocity 28 m/s. angle of attack 10°, $Re = 7.4 \times 10^4$

Fig.16.11 Flow around an automobile (smoke wire method)

16.2.6 Optical visualisation method

This method, whose most significant characteristic is the capability of complete visualisation without affecting the flow, is widely used. The Schlieren method utilises the change in diffraction rate due to the change in density (temperature). The interference method, which uses the fact that the number of interference fringes is proportional to the difference in density, is mostly applied to air flow. For free surface water flow, the stereophotography method is used. The unevenness of a liquid surface is stereophotographed to determine the difference in the height of the liquid surface and thus the state of flow is known. The moiré method is also used for water flows. The state of the flow is checked by obtaining as light and dark stripes the contours indicating the unevenness of the liquid surface.

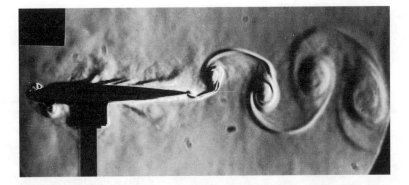

Fig.16.12 Flow at bottom dead point of vertically vibrating wing (Schlieren method) in air, flow velocity 5 m/s, chord length 100 mm, $Re = 3 \times 10^4$, vibration frequency 90 Hz, single amplitude 4 mm

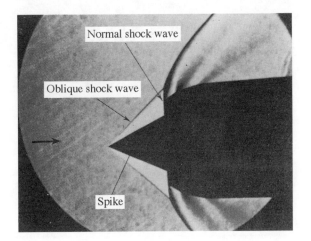

Fig.16.13 Flow at air inlet of supersonic aircraft engine (colour Schlieren method), $M = 2.0$, $Re = 1.0 \times 10^7$

A new technique, the laser holographic method, has been developed recently. An optical reference path is added to the optical system of the shadowgraph method or the Schlieren method.

Various actual examples of the optical visualisation method are shown in Plates 4[15] and 10[16] and Figs 16.12[17], 16.13[18] and 16.14.[19]

[15] Hara, N. and Yoshida, T., *Proc. of FLUCOME Tokyo '85*, Vol. II (1986), 725.

[16] Fujii, K., *Journal of Visualization*, 15 (1995), 142.

[17] Ohashi, H. and Ishikawa, N., *Journal of the ISME*, 74 (1975), 1500.

[18] Asanuma, T. *et al.*, Report of Aerospace Research Institute of University of Tokyo, 9 (1973), 499.

[19] Nagayama, T. and Adachi, T., *Joint Gas Turbine Congress*, Paper No. 36 (1977).

Fig.16.14 Equidensity interference fringe photograph of driven blade on low-pressure stage in steam turbine (Mach–Zehnder interferometer method) in air, inlet Mach number 0.275, outlet Mach number 2.123, pitch 20 mm

16.3 Computer-aided visualisation methods

16.3.1 Visualised image analysis

In this method, a visualised image is put into a still or video camera so that its density values are digitised. It is then put into a computer to be processed analytically, statistically, in colour distribution and otherwise, and thus is made much easier to interpret. Various techniques for this method have been developed. Among them, PIV (Particle Imaging Velocimetry) in particular has recently been popular. As an example of PTV (Particle Tracking Velocimetry), Plate 11 shows the velocity vectors obtained for flow over a cylinder by following, from time to time, the spherical plastic tracer particles of diameter 0.5 mm suspended in the water.[20] Plate 12 is an example of an image treated by a density correlation method. The image was obtained by injecting a smoke tracer into the room from the floor under the chair on which a man was sitting and natural convection around a human body was visualised.[21] Figure 16.15 is an example of the hydrogen bubble technique

---

[20] Boucher, R. F. and Kamala, M. A., *Atlas of Visualization*, Vol. 1 (1992), 197.
[21] Kobayashi, T. *et al.*, *Journal of Visualization*, 17 (1977), 38.

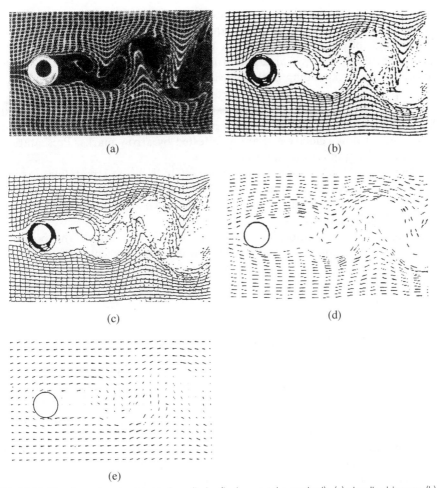

(a)

(b)

(c)

(d)

(e)

Fig.16.15 Kármán vortex street behind a cylinder (hydrogen tube method): (a) visualised image; (b) binarisation; (c) change to fine line; (d) velocity vector; (e) velocity vector at grid point

where the time line and the streak line are visualised simultaneously. The visualised image is caught by a CCD camera, converted to binary codes and fine lines, and thus the velocity vector is obtained.[22]

In Plate 4, the flow around a cone flying at supersonic speed is visualised by the laser holographic interferometer method, and the density distribution on a section is obtained by the computer tomography method.

16.3.2 Numerical data visualisation method

In this method, a flow field is numerically analysed by computer, and its

[22] Nakayama, Y. *et al.*, Report of Research Results (5th), Faculty of Engineering, Tokai University (1987), 1.

enormous computational output is presented in an easy-to-understand figure or image by computer graphics techniques.

The kinds of presentation include: contours, where physically equal values are connected by a curve; area colouring, where areas are painted in colours respectively corresponding to the physical quantity level of areas; isosurface, where physically equal values are three-dimensionally manifested in surfaces; volume rendering, where the levels expressed in isosurfaces are manifested by changing the degree of transparency; and vectorial, where sizes and directions of flow velocity etc. are manifested by arrow marks. Presentation can also be as graphs or animation.

Examples of contour presentation are Fig. 15.4, where streamlines (which are the contours of stream function) and contours of vorticity are manifested, Fig. 15.10, where contours of density are shown, and Plate 5 where the presentation is made three-dimensionally.

Examples of area colouring are Plates 1(a) and 2, where the pressure distribution is shown, and Plate 6(a) where the presentation is three-dimensional. And an example of isosurface presentation is shown in Plate 13,[23] and those of the vector presentation in Fig. 15.25(b), Fig. 15.26(b), Plate 1(b) and Plate 3.

16.3.3 Measured data visualisation

If a flow field is minutely measured with a Pitot tube, hot-wire anemometer, laser Doppler velocimeter, pressure gauge, thermometer, etc., such results can be processed by computer, and thus the phenomena are visualised as images.

In Plate 14, pressure-sensitive light-emitting diodes are placed transversely. The total pressure pattern of a wake of an aircraft wing is then obtained by photographing the diode emissions, whose colours change with total pressure.[24] Figure 16.16 shows the measured result of the flow velocity in the area behind a model passenger car obtained using a three-dimensional laser Doppler velocimeter, presented as a velocity vector diagram.[25] In Fig. 16.17 the acoustic power flow from a cello is visualised by the acoustic intensity method. The size and direction of the energy flow at each point is obtained through a computational process from the cross-vector of the sonic pressure signal on a microphone.[26]

[23] Miyachi, H., *How to Visualize your Data using AVS*, (1995), Fig. 5.28, Kubota Co., Tokyo.
[24] Visualization Society of Japan, *Fantasy of Flow*, (1993), 47, Ohmsha, Tokyo, and IOS Press, Amsterdam.
[25] Visualization Society of Japan, *Computer Graphics of Flow*, (1996), 124, Asakura Shoten, Tokyo.
[26] Tachibana, H. *et al.*, *Atlas of Visualization*, Vol. 2, (1996), 203.

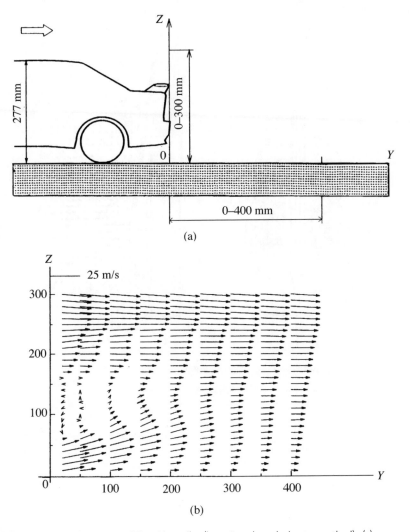

Fig.16.16 Flow behind an automobile with spoiler (laser Doppler velocimeter method): (a) measured region; (b) mean velocity vector

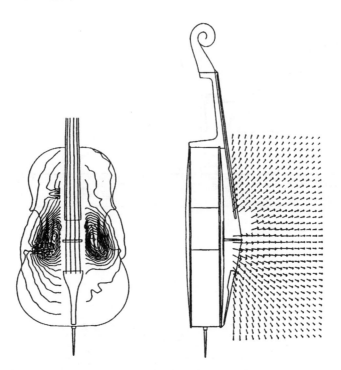

Fig.16.17 Radiating power flow of a cello (acoustic intensity method)

Answers to problems

2. Characteristics of fluids

1. $\mathrm{kg\,m/s^2}$

2. Viscosity: $\mathrm{Pa\,s}$, Kinematic viscosity: $\mathrm{m^2/s}$

3. $v = 0.001\,\mathrm{m^3/kg}$

4. $2.06 \times 10^7\,\mathrm{Pa}$

5. $h = \dfrac{2T\cos\theta}{\rho g b}$, $h = 1.48\,\mathrm{cm}$

6. $291\,\mathrm{Pa}$

7. $9.15 \times 10^{-3}\,\mathrm{N}$

8. $1.38\,\mathrm{N}$

9. $1461\,\mathrm{m/s}$

3. Fluid statics

1. $6.57 \times 10^7\,\mathrm{Pa}$

2. (a) $p = p_0 + \rho g H$,
   (b) $p = p_0 - \rho g H$,
   (c) $p = p_0 + \rho' g H' - \rho g H$

3. (a) $p_1 - p_2 = (\rho' - \rho)gH + \rho g H_1$,
   (b) $p_1 - p_2 = (\rho - \rho')gH$

4. $50\,\mathrm{mm}$

5. Total pressure $P = 9.56 \times 10^5\,\mathrm{N}$, $h_\mathrm{c} = 6.62\,\mathrm{m}$

6. 2.94×10^4 N, 5.87×10^4

7. 9.84×10^3 N

8. Force acting on the unit width: 1.28×10^6 N, Action point located along the wall from the water surface: 11.6 m

9. 7700 N m

10. Horizontal component $P_x = 1.65 \times 10^5$ N, Vertical component $P_y = 1.35 \times 10^5$ N, total pressure $P = 2.13 \times 10^5$ N, acting in the direction of $39.3°$ from a horizontal line

11. $976 \, \text{m}^3$

12. $h = 0.22$ m, $T = 0.55$ s

13. $\omega = \dfrac{1}{r_0}\sqrt{2gh'}$ rad/s, $\omega = 14$ rad/s at $h' = 10$ cm, speed of rotation when the cylinder bottom begins to appear $n = 4.23 \, \text{s}^{-1} = 254$ rpm

4. Fundamentals of flow

1. (a) A flow which does not change as time elapses is called a $\boxed{\text{steady}}$ flow. $\boxed{\text{Velocity}}$, $\boxed{\text{pressure}}$ and $\boxed{\text{density}}$ of flow in a steady flow are functions of position only, and most of the flows studied in hydrodynamics are steady flows. A flow which changes as time elapses is called an $\boxed{\text{unsteady}}$ flow. $\boxed{\text{Velocity}}$, $\boxed{\text{pressure}}$ and $\boxed{\text{density}}$ of flow in an unsteady flow are functions of $\boxed{\text{time}}$ and $\boxed{\text{position}}$. Flows such as when a valve is $\boxed{\text{opened}}$ / $\boxed{\text{closed}}$ or the $\boxed{\text{discharge}}$ from a tank belong to this flow.

   (b) The flow velocity is $\boxed{\text{proportional}}$ to the radius for a free vortex flow, and is $\boxed{\text{inversely}}$ $\boxed{\text{proportional}}$ to the radius for a forced vortex flow.

2. $\Gamma = 0.493 \, \text{m}^2/\text{s}$

3. $Re = 6 \times 10^4$, turbulent flow

4. $\dfrac{dx}{x} = -\dfrac{dy}{y}$ namely $xy = $ const

5. (a) Rotational flow
   (b) Irrotational flow
   (c) Irrotational flow

6. Water $v_c = 23.3\,\mathrm{cm/s}$, air $v_c = 3.5\,\mathrm{m/s}$

7. $\Gamma = 82\,\mathrm{m^2/s}$

5. One-dimensional flow

1. See text.

2. $v_1 = 6.79\,\mathrm{m/s}$, $v_2 = 4.02\,\mathrm{m/s}$, $v_3 = 1.70\,\mathrm{m/s}$

3. $p_2 = 39.5\,\mathrm{kPa}$, $p_3 = 46.1\,\mathrm{kPa}$

4. p_0: Atmospheric pressure, p: Pressure at the point of arbitrary radius r

$$p_0 - p = \frac{\rho Q^2}{8\pi^2 h^2}\left(\frac{1}{r^2} - \frac{1}{r_2^2}\right)$$

Total pressure (upward direction) $P = \dfrac{\rho Q^2}{4\pi h^2}\left[\log\dfrac{r_2}{r_1} - \dfrac{1}{2}\left(1 - \dfrac{r_1^2}{r_2^2}\right)\right]$

5. $v_r = 5.75\,\mathrm{m/s}$, $p_r - p_0 = -1.38 \times 10^4\,\mathrm{Pa}$

6. $t = \dfrac{2A\sqrt{H}}{Ca\sqrt{2g}}$

7. Condition of section shape $H = \left(\dfrac{\pi v}{Ca\sqrt{2g}}\right)^2 r^4$

   $Q = 12.9\,\mathrm{m^3/s}$, $d = 1.29\,\mathrm{mm}$

8. $H = 2.53\,\mathrm{m}$

9. $Q_1 = \dfrac{1 + \cos\theta}{2}Q$, $Q_2 = \dfrac{1 - \cos\theta}{2}Q$, $F = \rho Q v \sin\theta$

   $Q_1 = 0.09\,\mathrm{m^3/s}$, $Q_2 = 0.03\,\mathrm{m^3/s}$, $F = 2.53 \times 10^4\,\mathrm{N}$

10. $-7.49\,\mathrm{m H_2O}$

11. $n = 6.89\,\mathrm{s^{-1}} = 413\,\mathrm{rpm}$, torque $8.50 \times 10^{-2}\,\mathrm{N\,m}$

12. $F = 749\,\mathrm{N}$

6. Flow of viscous fluids

1. See text.

2. $\dfrac{1}{r}\dfrac{\partial(rv)}{\partial r} + \dfrac{\partial u}{\partial x} = 0$, or $\dfrac{\partial v}{\partial r} + \dfrac{v}{r} + \dfrac{\partial u}{\partial x} = 0$

3. (a) $u = 6v\left[\dfrac{y}{h} - \left(\dfrac{y}{h}\right)^2\right]$,

   (b) $v = \dfrac{1}{1.5}u_{max}$

(c) $Q = \dfrac{h^3}{12\mu}\dfrac{\Delta p}{l}$,

(d) $\Delta p = \dfrac{12\mu l Q}{h^3}$

4. (a) $u = 2v\left[1 - \left(\dfrac{r}{r_0}\right)^2\right]$

(b) $v = \dfrac{1}{2}u_{max}$

(c) $Q = \dfrac{\pi d^4}{128u}\dfrac{\Delta p}{l}$

(d) $\Delta p = \dfrac{128\mu l Q}{\pi d^4}$

5. (a) $v = 0.82u_{max}$,

(b) $r = 0.76r_0$

6. $\varepsilon = 4.57 \times 10^{-5}\,\mathrm{m^2/s}, l = 2.01\,\mathrm{cm}$

7. $Q = \dfrac{\pi d h^3}{12\mu}\dfrac{\Delta p}{l}$

8. $h_2 = 0.72\,\mathrm{mm}$

9. LT^{-1}

10. $8.16\,\mathrm{N}$

7. Flow in pipes

1, 2, 3, 4. See applicable texts.

5. See applicable text. Error of loss head h is 5α (%)

6. $h = 733\,\mathrm{m}$ at diameter 50 mm, $h = 26.4\,\mathrm{m}$ at diameter 100 mm

7. 24.6 kW

8. Pressure loss $\Delta p = 508\,\mathrm{Pa}$

9. $3.2\,\mathrm{cm}\,H_2O$

10. $h_s = 6.82\,\mathrm{cm}, \eta = 0.91$

8. Flow in water channel

1. $i = \dfrac{4.56}{1000}$

2. From Chézy's equation $Q=40.4\,\text{m}^3/\text{s}$, from Manning's equation $Q=40.9\,\text{m}^3/\text{s}$

3. $Q = 19.3\,\text{m}^3/\text{s}$

4. Flow velocity becomes maximum at $\theta = 257.5°$, $h = 2.44\,\text{m}$ and discharge becomes maximum at $\theta = 308°$, $h = 2.85\,\text{m}$

5. Tranquil flow, $E = 1.52\,\text{m}$

6. $h_c = 0.972\,\text{m}$, $3.09\,\text{m/s}$

7. $Q_{max} = 14.4\,\text{m}^3/\text{s}$

8. $1.18\,\text{m}$

9. See applicable text.

9. Drag and lift

1. Using Stokes equation, terminal velocity $v = \dfrac{d^2 g}{18v}\left(\dfrac{\rho_s}{\rho_w} - 1\right)$ where d is diameter of a spherical sand particle and ρ_w, ρ_s are density of water and sand respectively.

2. $D = 1450\,\text{N}$, Maximum bending moment $M_{max} = 3620\,\text{N m}$

3. $D = 2.70\,\text{N}$

4. $\delta_{max} = 3.2\,\text{cm}$ at wind velocity $4\,\text{km/h}$, $\delta_{max} = 4.1\,\text{cm}$ at wind velocity $120\,\text{km/h}$

5. $T = 722\,\text{N m}$, $L = 4.54 \times 10^4\,\text{N m/s}$

6, 7. See texts.

8. $D_f = 88.9\,\text{N}$, Required power $P = 133\,\text{N m/s}$

9. $L = 3.57\,\text{N}$

10. $D = 134\,\text{N}$

10. Dimensional analysis and law of similarity

1. Consider v, g, H as the physical influencing quantities and perform dimensional analysis. $v = C\sqrt{gH}$

2. $D = C\mu U d$

3. $a = C\sqrt{\dfrac{K}{\rho}}$

4. $D = \rho L^2 v^2 f\left(\dfrac{v}{\sqrt{Lg}}\right)$

5. $Q = C \dfrac{d^4}{\mu} \dfrac{\Delta p}{l}$

6. $\delta = x f \left(\dfrac{Ux}{v} \right)$

7. $C = f \left(\dfrac{d \sqrt{2\rho \Delta p}}{\mu} \right) = f(Re)$

8. (a) $167 \, \text{m/s}$

   (b) $33.3 \, \text{m/s}$

   (c) $11.1 \, \text{m/s}$

9. Towing velocity for the model $v_m = 2.88 \, \text{m/s}$

10. $\dfrac{1}{2.36}$

11. Measurement of flow velocity and flow rate

1. $v = 4.44 \, \text{m/s}$

2. $v = 28.5 \, \text{m/s}$

3. Mass flow rate $m = 0.325 \, \text{kg/s}$

4. $C_c = 0.64$, $C_v = 0.95$, $C = 0.61$

5, 6, 7. See applicable texts.

8. $U = 50 \, \text{cm/s}$

9. See applicable texts.

10. Error for rectangular weir is 3%, error for triangular is 5%.

12. Flow of ideal fluid

1. $\phi = u_0 x + v_0 y$, $\psi = u_0 y - v_0 x$

2. See applicable text.

3. Flow in counterclockwise rotary motion, $v_\theta = \Gamma / 2\pi r$, $v_r = 0$, around the origin.

4. $\phi = \dfrac{q}{2\pi} \log r$, $\psi = \dfrac{q}{2\pi} \theta$

5. Putting $r = r_0$, $\psi = 0$, the circumference becomes one stream line. Velocity distribution $v_\theta = -2U \sin \theta$, Pressure distribution $\dfrac{p - p_\infty}{\rho U^2 / 2} = 1 - 4 \sin^2 \theta$

6. The flow around a rectangular corner.

7. Flow in clockwise rotary motion, $v_\theta = -\dfrac{\Gamma}{2\pi r}$, $v_r = 0$, around the origin.

8. $w = Uze^{-ia}$

9.

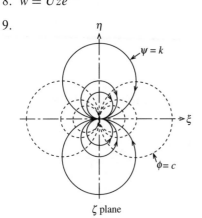

10.

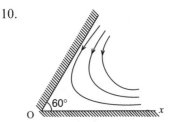

13. Flow of a compressible fluid

1. $\rho = \dfrac{p}{RT} = 1.226 \, \text{kg}/\text{m}^3$

2. $a = \sqrt{kRT} = 1297 \, \text{m}/\text{s}$

3. $T_2 = T_1 + \dfrac{1}{2}\dfrac{\kappa - 1}{\kappa}\dfrac{1}{R}(u_1^2 - u_2^2) = 418 \, \text{K}$

   $t_2 = 145°\text{C}$

   $p_2 = p_1\left(\dfrac{T_1}{T_2}\right)^{\kappa/(\kappa-1)} = 3.4 \times 10^5 \, \text{Pa}$

4. $T_0 = 278.2 \, \text{K}$, $t_0 = 5.1°\text{C}$

   $p_0 = 6.81 \times 10^4 \, \text{Pa}$

   $\rho_0 = 0.85 \, \text{kg}/\text{m}^3$

5. $v = 444 \, \text{m}/\text{s}$

6. $M = 0.73$, $a = \sqrt{\kappa RT} = 325 \, \text{m}/\text{s}$, $v = aM = 237 \, \text{m}/\text{s}$

7. $v = 272\,\text{m/s}$

8. $\dfrac{p}{p_0} = 0.45 < 0.528$, $m = 0.0154\,\text{kg/s}$

9. $\dfrac{A_2}{A_*} = 1.66$

10. $A_2 = 2354\,\text{cm}^2$

11. $2.35 \times 10^5\,\text{N}$

12. Mach number 0.58, flow velocity 246 m/s, pressure 2.25×10^5 Pa

14. Unsteady flow

1. $\dfrac{dz}{dt} = \pm 1.39\,\text{m/s}$, $T = 1.57\,\text{s}$

2. $T = 2\pi\sqrt{\dfrac{l}{g(\sin\theta_1 + \sin\theta_2)}}$

3. $0.69\,\text{m/s}$

4. $t = 1\,\text{min}\ 20\,\text{s}$

5. $a = 837\,\text{m/s}$

6. $\Delta p = 2.51 \times 10^6\,\text{Pa}$

7. $p_{\max} = 1.56 \times 10^6\,\text{Pa}$

Index